MY ODYSSEY

MY ODYSSEY

MEMOIRS OF THE MAN BEHIND THE 'MANGALYAAN' MISSION

K. RADHAKRISHNAN

WITH

NILANJAN ROUTH

PENGUIN

VIKING

An imprint of Penguin Random House

VIKING

USA | Canada | UK | Ireland | Australia
New Zealand | India | South Africa | China | Singapore

Viking is part of the Penguin Random House group of companies
whose addresses can be found at global.penguinrandomhouse.com

Published by Penguin Random House India Pvt. Ltd
4th Floor, Capital Tower 1, MG Road,
Gurugram 122 002, Haryana, India

First published in Viking by Penguin Random House 2016

10 9 8 7 6 5 4 3 2

ISBN 9780670089062

Typeset in Adobe Garamond Pro by Manipal Digital Systems, Manipal
Printed at Replika Press Pvt. Ltd, India

www.penguin.co.in

To the beloved members of my ISRO family—past,
present and future

CONTENTS

ACRONYMS

ASLV	Augmented Satellite Launch Vehicle
BARC	Bhabha Atomic Research Centre
CARE	Crew Module Atmospheric Re-entry Experiment
DOS	Department of Space
DOD	Department of Ocean Development
GAGAN	GPS Aided GEO Augmented Navigation
GOOS	Global Ocean Observing System
GSLV	Geosynchronous Satellite Launch Vehicle
HAL	Hindustan Aeronautics Limited
IAC	International Astronautical Congress
INCOIS	Indian National Centre for Ocean Information Services
ISAC	ISRO Satellite Centre
ISTRAC	ISRO Telemetry, Tracking and Command Network
IISc	Indian Institute of Science
IMSD	Integrated Mission for Sustainable Development
IOC	Intergovernmental Oceanographic Commission
IRS	Indian Remote Sensing
IIRS	Indian Institute of Remote Sensing
IMD	India Meteorological Department
LPSC	Liquid Propulsion Systems Centre
LVM3	Launch Vehicle Mark III (alias GSLV Mark III)
MCF	Master Control Facility

NNRMS	National Natural Resource Management System
NRSA	National Remote Sensing Agency
PSLV	Polar Satellite Launch Vehicle
RRSSCs	Regional Remote Sensing Service Centres
SAC	Space Applications Centre
SDSC SHAR	Satish Dhawan Space Centre at Sriharikota
SLV	Satellite Launch Vehicle
SSTC	Space Science and Technology Centre
TERLS	Thumba Equatorial Rocket Launching Station
VSSC	Vikram Sarabhai Space Centre

PREFACE

The first seven months of my second innings, starting from the dawn of 2015, were hectic yet inspirational. I delved deep into my passion and communicated with students, research scientists, corporate managers, entrepreneurs and the civil society through several forums on different occasions. My itinerary covered fifteen cities across the country, several other places in Bengaluru (earlier Bangalore) and my home state of Kerala. The themes on which I spoke were diverse. I intentionally chose distinct topics to expand and recapitulate my learnings. It also helped to break the monotony.

The most interesting thing was to identify a pattern in the questions posed to me during those interactions. During these sessions, the audience, despite their different backgrounds, asked about key success factors of ISRO. I consolidated my arguments and wrote a post titled 'Seven Reasons Why I Adore ISRO' for one of my social media pages. Later, a fairly long interview pertaining to my five-year tenure as the chairman of ISRO was published in a compendium, aptly titled 'From Fishing Hamlet to Red Planet— India's Space Journey'. The compendium covered all the glorious years of ISRO since its inception.

However, questions about my journey and experiences remained largely unanswered. That's when the idea to write my memoir seriously came about, and this project was born.

In April 2010, just within a fortnight of the failure of the Indian cryogenic upper stage on board GSLV-D3, Mr Y.S. Rajan, my nurturing boss of seven years during the 1980s, walked into my cabin and said, 'Radhakrishnan, you should start scribbling down the experiences of the past six months—what you had to go through to undertake this flight and how you faced the failure.' While the failure of the GSLV was a spicy starter, the rest of my five-year tenure turned out to be really exciting—I faced challenges, learnt a lot, evolved and went through several litmus tests of leadership. Meanwhile we, at ISRO, collectively achieved some great technological feats. One thing, however, remained consistent throughout—the bouquet of valuable life experiences.

In February 2010, I had handpicked Nilanjan from Antrix to handle several key and sensitive assignments in the scientific secretary's office at the ISRO headquarters. A young live wire with competence, character and candour, he would come up with fresh thoughts and ideas, which would, at times, clash with ours. I have beautiful memories of working with him at Antariksh Bhavan, where he often surpassed my expectations. A proficient colleague then and a close family friend now, Nilanjan had first-hand knowledge of my agonies and ecstasies of these eventful five years. Besides, he had a grasp of my life's experiences that I chose to share with him. In the penultimate days of my career, with all his youthful excitement, he suggested that I pen down my life lessons and also promised to help me whenever I decided to venture into it. I too was convinced that it was my duty to pen my experiences for the posterity. Nilanjan, thus, became my obvious choice to co-write this book.

I shared my intent to bring out this book with the chairman of ISRO, Mr A.S. Kiran Kumar, on 29 July 2015. He welcomed my resolve with spontaneity and graciously extended support with an office in the excellent ambience of Antariksh Bhavan to carry out this work. My affiliation with ISRO was thus refreshed in August 2015, after a self-imposed stint of detachment. Also, he concurred with my choice of the co-author.

Nilanjan and I ventured into our maiden attempt at writing soon. We got into the rhythm and completed the first seven chapters within a couple of months. The initial feedbacks from a few close friends were encouraging and we continued. However, we had to give it a month's break during November 2015 as each of us had other pressing commitments to attend to. This break helped us to critically review our work. Finally, we met our target of April 2016 and completed the manuscript. I could dedicate all my time to the book, but Nilanjan had to work on it as an add-on to his professional commitments at ISRO headquarters. I really appreciate his zeal and commitment to bring out the best in both of us through this work.

Since my early professional days, I inculcated the habit of making notes of whatever I grasped from important deliberations and events. It became more meticulous after my doctoral work. The spiral-bound notebooks that I had maintained (since 2000) became my prime references for chronicling my journey. I have often been complimented on my memory. The credit goes to these notes. Practice of Carnatic music enhanced it further. I owe a lot to these instruments.

One does not get an opportunity to write their memoir a second time. Thus, our intention was to articulate the life lessons and experiences in the best possible manner without hurting the sentiments of any of the actors in my life. At the same time, I could not afford to be dishonest to myself while narrating a couple of turning points and stormy events in my life. We have tried to capture them without embarrassing the people or the institutions. I sincerely hope the readers take the narration in the right spirit and focus more on the lessons that life offered me. After all, it all boils down to seeing the glass half full instead of half empty.

After completing this book, I have started assessing future scenarios of space exploration, space technology and space applications at a global level that would be relevant for the next few decades. My current engagements with ISRO and several academic institutions will enhance the academic content of this study. Besides this, my pursuit of Carnatic music continues, with regular lessons from Vidwan R.S. Ramakanth and occasional performances on renowned platforms. The most recent one was a seventy-five-minute performance on 31 August 2016 at the Chembai Vaidyanatha Bhagavathar Music Festival held at Sree Krishna Gana Sabha, Chennai. This was sponsored by the legendary Dr K. J. Yesudas. Thus, life goes on meaningfully.

ACKNOWLEDGEMENTS

I am grateful to thousands of individuals whom I have been acquainted with in my life journey of more than six decades. I have strived to learn something from every individual I have come across. This book itself is my tribute to everyone who helped me in my odyssey. I salute each one of them. However, I have taken the liberty to mention only a few of those names due to literary sensibleness. I hope to be forgiven by those not mentioned. Even though their names are not part of this narrative, they still remain fresh in my heart. Their warmth of love and affection is an asset, which I will always cherish.

I thank Mr Y.S. Rajan for triggering the thought process and my co-author Nilanjan for propelling me to make it happen.

A few of my close friends and associates went through the manuscript at various stages. I am indebted to the painstaking efforts of Mr Arun Balakrishnan (my classmate at IIM Bangalore and the former chairman and managing director of Hindustan Petroleum Corporation Ltd.); Mr V. Koteswara Rao, Mr S. Somanth, Dr B. Deependran and Mr T. Shamsudheen (representing four generations of close colleagues at ISRO); and Mr Rakesh Godhwani, a young faculty member of IIM Bangalore and a brilliant communicator, whom I have known for the past seven years. A special thanks is reserved for Mr P. Kunhikrishnan, director of the Satish Dhawan Space Centre, Sriharikota, for providing the photographs that speak more than several words.

I want to express my profound gratitude to Mr A.S. Kiran Kumar for supporting this work. Also, I appreciate that he read the manuscript and offered his excellent feedback.

My memoir is incomplete without my family and I express gratitude to all of them. Readers will meet all of them in the early pages of the book. Specifically, I wish to inscribe the emotional support that I gained from my beloved life partner Mini (Padmini K.V.), our endearing niece Dhanya, her admirable husband Dhruv and their little one Darsh for me to complete the manuscript of my maiden book within seven months.

Nilanjan's beloved family—Anindita and Abhijnan—were deprived of a significant chunk of his quality time due to his engagement with this book. I am grateful to them.

Mr Raghunatha Rao, my close associate at the headquarters from the mid-1980s, took care of me while I was engrossed in the manuscript. I thank him for his support.

I wish to thank Mr Udayan Mitra of Penguin Random House for his enthusiasm to publish this work and for assigning me an excellent team who made it possible to bring out this work in 2016. The editorial guidance and support from the brilliant line editor Richa Burman and the meticulous copy-editing by Saloni Mital deserve a special mention.

Finally, I wish to thank you, the reader, for taking interest in this memoir.

1

A GRACEFUL GOODBYE

The chill and fog of peak winter had gripped Delhi like a python grabs its prey. We were seated inside a well-furnished Air India Dreamliner; it was going to be a long wait as the Air Traffic Control (ATC) took more than two hours to allow the aircraft to start taxiing. It was 30 December 2014, and we were returning from Delhi after attending a day-long 'Make in India' workshop at Vigyan Bhavan the previous day.

We went straight to the office from Bengaluru airport, as we had already lost half the day. The rest of the day at Antariksh Bhavan in Bengaluru was not too eventful. I had already decided that in my last month at the helm of affairs of ISRO, I wouldn't take any major decision that could, perhaps, tie down the hands of my successor.

I returned home from the office at the usual time and changed into my comfortable house wear. With the sublime background of a light classical number, I started chit-chatting with my wife, Mini. This had been the custom for more than three decades.

At the dinner table, Mini suddenly asked me, 'Do you realize what the date is tomorrow?' I caught a breath between bites and said, 'Why not! I am looking forward to handing over the baton tomorrow.'

We hardly ever talked about official matters at home, sparing some issues which impacted me personally. But my penultimate day in office was too special to be confined within that bracket.

I paused and said, 'I did reiterate my recommendation for succession two months ago. Apparently the choice is yet to be made. We should know by tomorrow. And of course I would love to have a "thank you" call from them. That's all.'

She quizzed further, 'How do you feel now?' I responded promptly, 'Blissful! Could I have asked for more than what we have done during the last five years? ISRO is now admired all over the country and abroad. We have completed thirty-seven missions; six of them in the last six months. We didn't have a great start, unnecessary turmoil consumed a lot of our energy. But we held our ground. The Geosynchronous Satellite Launch Vehicle (GSLV) with our own cryo stage is through; LVM3 made a fantastic debut and, of course, we reached Mars. I am leaving ISRO in great shape for my successor. I hope that posterity judges it that way.'

I took a deep breath in. I had been waiting for the right time to share these thoughts with Mini.

Mini probed me further, 'How was your Delhi trip?' I replied, 'Excellent, though the weather wasn't great. The day was full of meetings. I had to make two presentations—one to industry leaders and the other to the prime minister—in the valedictory session. The who's who of the government and industry were present in the packed plenary hall of Vigyan Bhavan. The prime minister looked satisfied with the report card of ISRO.'

'So, what's on your calendar tomorrow?' she asked. 'Well, it would be just another day at the office. In the evening I have to attend a function at the Press Club of Bangalore—they are honouring me with the "Man of The Year" award. The chief minister himself would do the honours. I will come home directly from the Press Club.'

She smiled; she was convinced that I was in the correct frame of mind for the D-Day.

~

The next morning, business started as usual. I set out at 9 a.m. for Antariksh Bhavan. My driver, Chandrasekhar, who had been with me for the entire tenure as chairman, was looking a bit anxious for a change. It is a short drive of ten minutes from 'Vyoma', my official residence in RT Nagar, to Antariksh Bhavan. Thankfully, we did not encounter much of the dreaded Bengaluru traffic en route.

In the office, my staff, Yashoda, Raghu, Sreedhar and Padmanabha, greeted me as usual, with beaming faces, and that marked a good beginning to the day.

A farewell function was arranged for the superannuating colleagues in the forenoon. Normally this happens unswervingly on the last working day of every month at all the offices of ISRO with a lot of warmth and affection, and then business continues as usual. A few of my old colleagues, including the then scientific secretary Prahlad Rao, were being felicitated at Antariksh Bhavan. But I excused myself from attending this function, as I was awaiting communication from Delhi about my successor.

As hours passed, the tension of uncertainty gripped the organization about the transition; nobody knew who the next leader would be. Transitions in ISRO's leadership had always been seamless, with the announcement of the successor being made well in advance. But it was not so in this case, even though A.S. Kiran Kumar, the director of the Space Applications Centre, Ahmedabad, had been inducted into the Space Commission six months ago.

Around noon, I called up Delhi and reminded them that this was my last day in office. Delhi said that it might take a couple of days more for the decision on the successor to come through. They suggested that I hand over additional charge to an interim. A few options were discussed. They said that they will confirm after the approval!

I walked into the chamber of Sudarsan Srinivasan, member (finance) of the Space Commission, and called in the additional secretary A. Vijay Anand, the joint secretary S. Kumaraswamy and the associate scientific secretary H.N. Madhusudhana to inform them about the developments.

I felt my 16,000-strong ISRO family deserved to know about my demitting office that evening, even though the official communication was yet to arrive. I requested Madhusudhana to arrange for a video address in the afternoon.

Soon, I sent word for Nilanjan to come to my room. He remarked, 'All faces in this building are gloomy, and you are sitting with such glee on your face!' As usual, we had lunch together in my anteroom, discussing nuances of my last address to the ISRO community as its chairman.

Koteswara Rao, my trusted lieutenant for more than two years, who had superannuated just two months ago as scientific secretary, walked in too and greeted me saying, 'I just wanted to be with you on this day.' Dr B.N. Suresh, who had known me since I joined ISRO, walked in too. I announced, 'I am joining your club of veterans from tomorrow.'

I felt I should seek the blessing of ISRO's 'Pitamah Bhishma' Prof. U.R. Rao, who had been a strong pillar in my career for the last three decades. I met him in his cabin at Antariksh Bhavan and requested his presence at my address.

The auditorium at Antariksh Bhavan was tightly packed. Emotions were visible on all faces as guesswork had reached its climax.

My town hall sessions, with my colleagues, were almost a quarterly affair. This used to be direct from the headquarters or one of ISRO's locations over video for teams at other locations.

I chose to be brief on that day, and to congratulate my most beloved organization on its stupendous achievements, I added:

This is the time to think of the great founders and thank them for the vision and culture they have passed on to us.

As an engineer, who joined ISRO in the Space Science and Technology Centre in May 1971, it has been a long journey for me. Today is the proudest moment of my career and life.

The future of ISRO is in your hands and I am sure that the future leadership of ISRO will take it to greater and greater heights and participate in the process of bringing India to the centre stage, the forefront in the world.

We will make our presence felt, not only on earth, but we will also reach the moon and Mars in the coming decades.

I wish all of you a happy and prosperous New Year. I want to thank you for all the support you have given me, especially during my difficult times in the last five years. Thank you very much.

I had not been explicit about this day being my last one in office. The reason was obvious. Communication from Delhi assigning additional charge to Shailesh Nayak reached Bengaluru only when we had settled for a cup of tea after the address.

I signed off and left the office, accompanied by Nilanjan, one last time in my capacity as chairman, ISRO.

I reached the Press Club of Bangalore in time. The chief minister of Karnataka was present, along with a few of his cabinet colleagues. The Press Club was packed with members of the media and rightly so. I excused myself from answering the obvious questions.

On the way back, Nilanjan showed me the farewell message that ISRO's social media team had put up for their 'Man of Steel'. I felt really great. The social media team did not have to seek approval from ISRO's top management for the content they posted online. We trusted them. What made it even more special was the way it was received by the netizens of the country. This was a beautiful gesture.

Some of my young and not so young colleagues from Antariksh Bhavan—Koteswara Rao, Kanungo, Shantanu, Gowrisankar, Yashoda, Raghu, Sreedhar, Ramachandra and Vani reached home with sweets and flowers before us. They joyfully declared, 'We are not here to bid you adieu, we are here to celebrate your success as the chief of ISRO. This is just a reminder that we will trouble you at your house whenever we want.'

We saw them off. Mini said, 'Nilanjan, stay back for a while, Dhanya and Dhruv are on their way. I will make dosas [rice crepes] for all of you.'

Dhanya, our niece, and her husband, Dhruv, were on their way from their respective offices in east Bengaluru, so we knew it would take time.

We settled around the round table, and Nilanjan gave a thumbnail sketch of the day's happenings to Mini.

She initiated, 'So, we are at the dawn of a new phase of life.'

I smiled and said, 'It was quite a journey for the last forty-three-odd years, a bit incredible if you ask me. I was just another schoolboy like the millions you meet in this country. My physical fitness for college education was in doubt. I strived hard and got into ISRO at Thumba as an electrical engineer. Dr S.C. Gupta mentored me. He also allowed me to pursue higher education in the discipline of management. In 1981, the chairman, Prof. Satish Dhawan, handpicked me for the ISRO headquarters, and I bloomed under his shadow. His successor, Prof. U.R. Rao, groomed me further for a decade, trusting me with several key roles.

'In 1997, I relocated to the National Remote Sensing Agency (NRSA) at Hyderabad to contribute and grow in a major centre of ISRO. It was a rough patch of three years, but it enhanced my emotional maturity. Meanwhile, I got a doctorate, under the guidance of Prof. Dhawan; I was his last student. My future still looked bleak and I chose to move out of ISRO.

'In July 2000, I had a professional rebirth in the realm of ocean sciences. I evolved into K.R. version 2.0 and excelled with some level of global visibility. A turn of events in 2005 reunited me with the ISRO family. The goodwill and mutual affection were evident. I headed NRSA for two years and then I was appointed to head VSSC, the largest centre of ISRO. And then I landed here in November 2009.

'And after these eventful last five years here, I feel happy and contented. Posterity might remember me for "the Mangalyaan" (Mars Orbiter Mission). But I would like to be remembered as a person who served the Indian space programme to the best of his abilities.

'After all this, today was supposed to be an emotionally difficult day, but I was mentally prepared for it after putting up the succession plan for the entire ISRO. The practice of amalgamating the attachment and detachment is indeed difficult but laudable at all levels. This is what I intend to do from tomorrow.'

On hearing this, they smiled.

~

Shailesh Nayak, a former member of the ISRO team and the then secretary of the Ministry of Earth Sciences, came to Bengaluru the next day to hold the fort for a short duration. His obvious enthusiasm was clearly visible as we sat together over a cup of lemon tea in my house before he took over. It was sheer coincidence that earlier he had succeeded me at the helm of the Indian National Centre for Ocean Information Services and Tsunami Warning Centre, as I had returned to the ISRO family in November 2005.

At last, on the evening of 12 January 2015, news of the appointment of Kiran Kumar as chief of the Indian space programme, in all its three key roles, came in. This was a moment of great happiness for me as I had strongly advocated this for months.

In Kiran, ISRO got the best chief to lead it to the next level of excellence. This man had played a pivotal role in the Mars Orbiter Mission (MOM), and also commanded exceptional professional respect and admiration from all sections of ISRO. He had evolved, at a pace beyond imagination, to become an effective vice chairman of the ISRO council, and had hugely contributed as a member of the space commission over the past six months.

Thankfully enough, ISRO was yet again set in the right direction.

2

TALE OF THE BOY NEXT DOOR

An invitation from the St Thomas College in Thrissur landed on my table one afternoon. The Department of Physics, which was set up in 1922, was celebrating its jubilee. I was invited to give a keynote address in the very classroom where my father, the late K.P. Krishnan Kutty Menon, studied physics at the intermediate level, in the early 1930s.

The timing of the invite was important as ISRO had just perfectly executed the PSLV-C15/Cartosat-2B mission (12 June 2010)—the first success after I took charge. This was after the debacle of the GSLV-D3 mission on 15 April of the same year. I was excited to go back to experience the sights and sounds of Thrissur, my home town, and easily the cultural nerve centre of Kerala.

We had a technical review at the Vikram Sarabhai Space Centre (VSSC) the previous day. I requested the then scientific secretary V.S. Hegde, B. Deependran, a brilliant aerospace engineer from VSSC, and Nilanjan to accompany me in this journey down memory lane. I wanted to show them around my cradle—Irinjalakuda, an ancient town where culture and communal harmony converge with great effect.

The address went well with the extremely bright audience of St Thomas College. I say so because the audience chose to reciprocate with sharp and intriguing questions on the Indian space programme rather than just applaud. I have always thought interacting with students is one of the most gratifying experiences; this event reaffirmed my belief.

Revisiting My Roots

After the function, we drove down to Irinjalakuda, just twenty kilometres away. Both Mini and I have ancestral houses here. At my ancestral house, 'Koppillil', we met my octogenarian aunt and sought her blessings. Even at this age, she has a lot of interest in my career and in my organization. I have always made a point to call her up before and after each mission, informing her about the details in a language that she can decipher.

We also strolled along the holy precincts of the Koodalmanikyam temple, the imposing traditional auditorium for Sanskrit theatre performances, the adjoining Kathakali school, Mahatma Gandhi Library and Reading Room, Mahatma Gandhi park, the Koodiyattam school, and the National High School, where I did my entire schooling, all within ten minutes from my ancestral house.

We settled in the P.W.D. Rest House, adjoining the school, ordered coffee and started talking. 'So this is where the chairman of ISRO grew up,' quipped Nilanjan. 'No', I replied. 'This is where Radhakrishnan, just another boy from Irinjalakuda, grew up.'

Hegde and I had known each other for two decades, yet he did not know much about my childhood. 'Sir, tell us more,' he said in his rich voice.

I loved narrating my story and continued enthusiastically, 'I hail from a typical matrilineal Nair joint family of Kerala. My maternal grandparents were teachers in the Government Boys High School in Irinjalakuda. After my grandfather's demise in 1960, my grandmother and her sister took charge of managing paddy and other cultivation on our joint property of nearly twenty acres. They also supervised the cattle herd and nearly twenty-five farm labourers.

'My grandfather would insist that along with education, the younger generation in the family should be conversant with all facets of farming and animal husbandry. I followed his instructions with eagerness and gained a reasonable level of proficiency in these areas. I also picked up a bit of carpentry and masonry.

'More importantly, I mingled with the farm labourers during their toil, and closely observed their arduous life in the tiny huts surrounded by paddy fields. They were invited to my birthday feast that falls in the harvest season, just a week before Onam. Even now, I keep in touch with them. I met one of them, a woman named Kunjikkaali of my mother's age, at our house last month. I was here to attend a civic function organized to felicitate me in the gracious presence of Dr A.P.J. Abdul Kalam and 5000 schoolchildren. You know, her three sons were my contemporaries and playmates.'

Nilanjan observed, 'That is why you are at ease with people from all kinds of backgrounds.'

I continued, 'My father graduated in physics from Maharaja's College in Ernakulam, and after that joined Kerala's revenue service. I have keenly observed him at work, especially as a land tribunal officer and as the deputy collector responsible for land acquisition for HMT Ltd and Cochin Oil Refineries in the 1960s. He was a diligent officer with high self-esteem. He was honest and had an unending willingness to help the needy and deserving. He was an avid reader of English novels. He enjoyed playing cards; this kept him mentally alert and honed his incisive, analytical capabilities.

'My mother, the late Koppillil Ammini Amma, studied mathematics at the Maharaja's College in Ernakulam. Later, she took up teaching in the government service. She was a soft-spoken, strong-willed and selfless person. She was also an active member in the Bharat Scouts and Guides movement. She retired as district educational officer, Kottayam, in 1978.

'I was the second child, younger than my brother Sivadasan by a year and older than my sister Radha by five years. My parents had set very high moral standards for all three siblings, constantly reminding us to admire the smaller things in life rather than complaining about shortcomings. Those were lessons for a life-time.

'My brother, Sivadasan, and I had contrasting personalities. He was appreciated by all in the family and school. He retired from VSSC in 2008, with the reputation of a quiet and extremely hard-working scientist. He is now settled at Thiruvananthapuram along with his wife, Usha. Even now our conversations over the telephone or otherwise do not last long as he prefers being a silent listener and observer. But he gives me his insights on some matter or the other, and I really treasure these. Even when we were in our forties, my brother was more intimate with our father, and I was close to our mother. This is why, probably, his advice was as important as my father's.

'My sister Radha has always been vibrant and talkative. A graduate in pharmacy, she kept everyone linked in the family. She spent most of her life in the northern and eastern parts of the country, where her engineer husband, P.N. Ramachandran, was engaged. She became a vital link between the various members of the erstwhile joint family. Their daughter, Maya Ramachandran, a brilliant engineer, is now with NTPC Limited in Faridabad.'

'Radhakrishnan and Radha! Is there a story behind this?' Nilanjan whispered. I explained, 'Yes, there is. My mother had a younger sister named Radha, a very bright student of physics. She died because of typhoid just before her final year BSc examinations in 1946. To commemorate her, both of us were named after her. Her books, notes and answer sheets, with compliments from professors, were all preserved. I got inspired by these, and aspired to live up to this name bestowed on me.'

I continued, 'The grand old lady whom you just met is my mother's first cousin, Kallianikutty Amma, fondly called Kochuvava by all. She was a mathematics teacher and retired as headmistress of the National High School. She married my elder maternal uncle, the late Balakrishna Menon, who was a manager in the Bank of Baroda. My second uncle, the late Jayaraman, another mathematics teacher, retired as headmaster. His wife, the late Jaya, was a science teacher at the Model Boys High School, Thrissur.'

Deependran commented, 'An amalgam of education and agriculture in a joint family.' I agreed.

School Life

We were so immersed in our conversation that we forgot about our return flight to Bengaluru, which we had to catch from Kochi airport—a two-hour drive from Irinjalakuda. We headed towards the airport.

We took a minor detour on the way and stopped at the entrance of the imposing edifice of Christ College, set up in 1956. I said, 'This academic tower, which is visible from my home, kindled the pursuit for higher education in me since childhood. I did two years of pre-degree, from 1964–66, in this college.'

We reached the airport and as luck would have it we found that the flight had been delayed. So we had some more time to extend our conversation, but this time at the cosy lounge of Kochi airport and over some piping hot coffee. Nilanjan initiated again, 'This college was just two kilometres from your home. Did you walk down or cycle?'

I smiled and said, 'I travelled by car, a Standard Herald. It was a sheer necessity then and not a luxury.' The three were puzzled.

I clarified, 'The year of 1964 greeted me with rheumatic arthritis that "licked my heart and bit my joints". I was incapacitated for two years; my knee joints remained affected, though mildly, for a few more years.

'But, in the process, I became a voracious reader. Our neighbour, Prof. Mampuzha K. Kumara Menon, a literary scholar and orator, helped me form a world view. Two articles that I wrote then—on special theory of relativity and nuclear power—were published in the college magazines. Also, I escaped the distractions of adolescence, just as I missed many other excitements.

'The disease finally disappeared in 1972 after biting my hip joints.'

Hegde commented, 'That was a blessing in disguise.' This time Deependran intervened, 'Sir, you showed us your school but tell us more about your time as a schoolgoing kid.'

It made me nostalgic. I recollected, 'It was a Malayam-medium school then. I was a very shy boy and the shortest in my class. Many who knew me in those days, still wonder how I evolved to this level, and how I managed to change my body language so much.

'In 1954, I stepped into that school as a student of class one. The primary wing was a bit away from the main building, housed in a thatched building. Our neighbour, the late Thankamani Amma, was the class teacher. She used to take me to school; my hand wrapped around hers. I had not attained the prescribed age of five years at the time of joining. Hence I was a "bona fide student", who would be considered for regular admission a year later, in class two, based on merit. There were no roll call or mark list for "bona fide students" like me.

'There is an interesting anecdote. Once, I noticed that my mother had taken casual leave and I wanted to stay back with her. I started crying in the classroom citing "stomach ache"; it worked well that day. After that, I frequently began to complain of stomach ache even while getting ready for school. Soon the cat was out of the bag and I got the nickname "stomach ache". I knew I couldn't use the excuse any more.

'I ranked second in the Kerala state-level merit scholarship that one is eligible for after completing primary schooling. When the news came, my mother commented, "The credit should go to Ambujam who took the pain of coaching him and ten others voluntarily." To some, this may be an illustration of the teacher–parent relation that existed in those days. But from my mother's comment, I learnt a valuable lesson: credit should go where it is due.

'Our school is well known for high standards of pedagogy and the results that the students achieve year after year. Of course, we had excellent teachers and some of them are still alive. I make it a point to meet them whenever I get a chance.

'The school campus used to reverberate with several activities such as fine arts, sports, games, social service, National Cadet Corps, scouts, career counselling and so on. Fine arts became my forte. One of my classmates, Mr Innocent, film artiste and currently a member of Parliament in the Lok Sabha, recounts this even now. I had my first stage show right there. I performed a folk dance at the age of seven.'

My audience looked a bit amazed. We were about to order a second round of coffee when we heard the boarding call. We bid adieu to Deependran who had to start for his home town to spend an evening with his parents.

I boarded the aircraft feeling rejuvenated; it was a day well spent.

3

TRANSFORMATION FOR THUMBA

'You go to Thumba. They have plans to develop the Satellite Launch Vehicle (SLV). It would be wonderful for the country if they are able to do it,' asserted Prof. M.S. Narayanan Potti. He was a brilliant academician and an immensely caring mentor at the Government Engineering College, Thrissur (GECT). I would consult him whenever I faced any dilemma. This time it was a tough one—whether to attend the recruitment process at the International Business Machines Corporation (IBM) or at the Space Science and Technology Centre (SSTC) at Thumba (which later transformed to the Vikram Sarabhai Space Centre or VSSC), scheduled on the same day.

This situation of 'problem of plenty' came after an agonizing six–seven months. By the time our results for engineering final exams were announced, in August 1970, our batch had missed the deadline for admission to other institutes for higher education in that academic year. I started looking for avenues for higher education or a befitting job opportunity. I learnt a very important lesson that time—on how to face odds and obstacles in life and still stay positive to create better opportunities. That lesson helped me on several occasions in the years to come.

I wrote an entrance test (there was no CAT then) for admission to the PG diploma in management course at IIM Calcutta in their eighth batch. Even though I cleared the test, I couldn't complete the process as the second phase of interviews and group discussion was coinciding with the commencement of a one-year training at Premier Tyres, Kalamassery. I opted for the latter as it is said 'a bird in the hand is worth two in the bush'. I was settling into the groove of the demanding and gruelling training in the plant engineering department when the next predicament showed up—that of choosing Thumba or IBM.

Prof. Potti's words were a godsend, and I felt I did not need anyone else's advice on this matter.

On the morning of 19 April 1971, I arrived at the recruitment office of SSTC to attend the interview for the post of Technical Assistant-C. At that time, this was the entry-level post for an engineer, who had graduated with first class or better. The picturesque Veli Hills were not very well connected through public transport. I was familiar with the route, as I had visited the adjoining Thumba Equatorial Rocket Launching Station (TERLS) at the end of my engineering course. The competition was tough, I felt, as four more candidates had turned up that day. Among them, two were from Jadavpur University, one was from Coimbatore and the other from Kerala.

Soon we were told that the chairman of the selection committee was busy and the recruitment interview would commence only by 3.30 p.m. There was ample time. I settled in an armless chair to recapitulate and brush up my preparations for the interview. I went through a mental simulation of coining relevant questions and sharpening the answers. I felt confident and comfortable to face the interviewers.

Two of my intimate friends and classmates at GECT—N.S. Vidyasagar and Josekutty Joseph—were already working with the SSTC for the past seven months in the Propulsion Division. They showed up and wished me luck. They promised to come back and take me for lunch after the interview.

This meeting brought back memories of my engineering college days. A degree in engineering was not my first choice. I always aspired to study mathematics, having secured 99 per cent in class ten and 100 per cent in pre-degree. I was ecstatic to join Maharaja's College, Ernakulam, for the special BSc course in mathematics. Soon, we came to know that I had been selected for the BSc (Engg) course at the Regional Engineering College (REC), Kozhikode. I did not want to take it. The reluctant adolescent in me was finally persuaded by my resolute second uncle, Jayaraman, to switch over to engineering for better prospects. I agreed hesitantly.

I was packed in the same old Standard Herald and driven to REC, Kozhikode, under the strict vigil of my father and uncle, Balakrishna Menon. Homesickness kept haunting me there as 'interactions' with seniors in the hostel got more intense. I found occasional solace in the company of two faculty members who were my mother's students in school. A week passed without the classes commencing.

One fine morning, the local newspaper published the list of students who had been selected for the Government Engineering Colleges at Thiruvananthapuram and Thrissur. Both of these were affiliated to the Kerala University back then. I was thrilled to see that my name was the fifth one on the list for Thrissur.

Homesickness prevailed over everything else, and I ran away from the REC hostel in the forenoon of the same day. That one week in the hostel had made me so bold that I decided to undertake my first solo trip. I walked for nearly five kilometres and took three different buses to cover the 140 kilometres to my home; I reached in the wee hours of the next morning.

My astonished parents guessed the gravity of the situation. They readily agreed when I told them that I wished to study at the Government Engineering College, Thrissur. I was assigned to my second cousin E. Ravi, a final year BSc (Engg) student at GECT, for my guided entry and stay there.

It was election time for the students' council at GECT. So the new students were given royal treatment. The admission procedure was a breeze, and we were greeted with plenty of good wishes, promises and most importantly with leaflets that contained information about the candidates contesting the election. But to my surprise, after the elections interactions with seniors were conducted with due dignity. This cemented a strong and long-lasting bond between the seniors and juniors.

The sprawling seventy-five-acre campus of the decade-old GECT was situated in Ramavarmapuram, just five kilometres to the north of the famous Thekkinkadu Maidan, Thrissur. All India Radio, a Central Reserve Police Force camp and Vimala College were the other establishments in the suburb. To the north of the campus, there was a well-formed straight road of one kilometre that led to the state highway; this connected Thrissur and Shoranur.

Life at GECT triggered the process of a multifaceted transformation in me. The days were full of activities—classroom learning, interaction at the amphitheatres, workshops, electronics club and study tours—and the students were mentored by caring teachers. Those days, the best used to opt for the teaching profession and we students were the fortunate beneficiaries. There were stimulated interactions outside the classrooms, cutting across disciplines and batches. This broadened the horizons of students. As most of us stayed in the hostel, support from helpful senior hostel mates also contributed to the transformation.

In the first year, I secured first class, just enough to retain the national merit-based scholarship for teachers' children that was awarded to me in 1966. More importantly, I was allotted electrical engineering, my first preference of discipline.

That was when I came across assistant professor M.S. Narayanan Potti, my mentor-to-be, for the very first time. He taught us Electric Circuit Theory that year. A scholarly teacher, his conviction was contagious. He instilled great confidence and enthusiasm in me to rise to the top and become the best student in the batch throughout.

Rapt attention in the classroom and a good memory helped me to cope with the demanding coursework, and allowed me to find time for extracurricular pursuits. I became an active member of the fine arts club, drawn strongly towards dramatics. I acted in as many as nine dramas. A minor interest in badminton promptly vanished when arthritis threatened to come back.

In the evenings, a few of us would visit the nearby Manalarkavu Devi temple, while others opted for the exciting environs of Ramavarmapuram.

It was an emotional departure from GECT in May 1970. In the final examination, I secured first class with honours. I secured the first position in the college for BSc (Engg) in electrical engineering.

The head of the department, Prof. M.S. Abdul Khader, wrote in his testimonial, 'I am impressed by his clear ideas about theories of electrical engineering. He has a very good aptitude for research.' I preserved the testimonial as my greatest treasure ever.

~

'Not hungry?' Vidyasagar woke me up from the spell of remembrance and recollection. He and Josekutty had turned up as promised to take me for lunch at the office canteen.

'Thank you all,' is all I could say as a tribute to all those who helped that boy reach the portals of Thumba.

Finally, the selection committee assembled at 3.30 p.m. and Mr N. Sridharan Dhas chaired the interaction. The questions were mostly on electrical machines, control systems, my project work and fundamentals of digital electronics. I thought I did well.

After the interviews, we were waiting for directions to go for the medical test. I was called inside again and Mr Dhas asked me, 'How much time do you require to join?' I was excited and said, 'Fifteen days.' Mr Dhas retorted fiercely, 'You have stated in your application that you can join within two days.' I could not stop a smile.

Mr Dhas commanded, 'Okay, I am going to NASA for two months. You meet Embranthiri (Mahadevan) and study the book on Permanent Magnet Design and Applications.'

I could guess what was in the offing!

4

A THUMBA ENGINEER IN ACTION

Every individual has a day in their life that remains vivid as ever, and I am no exception. Even today when I sit back in remembrance, I relive each moment of that day. Tuesday, 4 May 1971, was when I first stepped through the portals of SSTC with my letter of appointment as a Technical Assistant-C.

My heart was throbbing with excitement. Thumba caught the imagination of all my contemporaries in those days. It was perceived as a wonderland of modern technology and quickly became a sought-after destination, only next to the famed Bhabha Atomic Research Centre (BARC). Of course, SSTC and BARC shared a kind of umbilical connection between them.

My day of joining was full of procedural formalities. And then, I was guided to the office of Dr S.C. Gupta located on the second floor of the main building at Veli Hills. Dr Gupta was heading the Control, Guidance and Instrumentation (CGI) division. He was away on tour for a week. Mr Dhas, who chaired my recruitment interview, used to report to Dr Gupta. He was yet to return from NASA.

Radhakrishnan Nair, a very able and amicable stenographer at the division, introduced me to the deputy head, Dr V.P. Kulkarni, who in turn suggested that I occupy a seat in the 'gyro laboratory' for the time being, as a seat was readily available only there.

That was quite a premium office space within the fully air-conditioned main building. The second floor of the main building hosted the heads of all divisions. The 'instrumentation laboratory', the 'gyro laboratory' and the control analysis engineers of CGI were located in three adjacent rooms on the second floor. On the opposite side, two rooms housed the laboratories of the Satellite Systems Division that later evolved into ISRO Satellite Centre (ISAC) in Bengaluru, under the dynamic leadership of Prof. U.R. Rao.

The induction was completed with these introductions, and I realized that my first day at ISRO was over. I was putting up at my brother Sivadasan's place, near the East Fort area of Trivandrum (now Thiruvananthapuram). We managed for a month there, before settling down at Chandragiri Lodge near the Statue junction. The lodge had nothing great to talk about, except for the fifty-odd occupants, who were a fine blend of faculty members of the engineering college, engineers from the state service, and about a dozen Thumba engineers. I also met some teachers and classmates from my college.

Learning on the Job

The next morning at the office, I saw Mr N. Vedachalam dashing into the gyro lab with the order letter of his out-of-turn promotion, to the post of Engineer-SD. I gathered that a few weeks ago, Dr Gupta's team had been successful in developing a rate gyro and that Mr Vedachalam was the key mechanical engineer of this team. My maiden lesson was on the rate gyro—an electromechanical instrument with a spinning rotor attached to a gimbal frame, it could measure the rate of angular motion of a flying rocket. Curiosity drove me to learn more about gyros and accelerometers. These tiny instruments work on properties of inertia, and hence are known as inertial sensors. They are required on a rocket launcher to estimate its position, velocity and orientation during flight, without any signal from the ground. Obviously, the rate gyro was a crucial milestone in the development of inertial sensors in the country at that time.

I took time to explore the buildings and facilities of SSTC and its environs. Vidyasagar and Josekutty rendered their guidance. The visual setting at Veli Hills was so invigorating. The vast Arabian sea, only a few metres away; traditional fishing near the coast using catamarans; ships disappearing over the horizon; Avro aircraft taking off from the adjacent airport; and meter-gauge rail tracks just a stone's throw away—all this contributed to the picturesque setting.

Formal induction programmes or initial training were not in vogue then. Everyone was expected to 'learn on the job'. SSTC had more than 1600 employees, mostly in their twenties. There were a few exceptions—mostly ex-servicemen or senior officials in their thirties.

The 'Design Project-Satellite Launch Vehicle [DP-SLV]' was the main priority with several of the seniors leading the definition and detailing of specific stages of the rocket or the subsystems. One could easily spot a group of well-groomed, buzzing, bright engineers at the centre, making beelines for the SSTC library. The library boasted an excellent collection of books and periodicals.

I soon realized that I needed to study extensively to survive and strive.

The scientists and engineers of Thumba, mostly bachelors, were obsessed with office work. I became a regular at the after-office discussions at Chandragiri Lodge with my contemporaries from diverse disciplines. This drove me to be a voracious reader. With all humility, I could say that most of the books at the then SSTC library were browsed by me; many of which I read, some of which I studied. This came in handy during my post-IIM years.

As days progressed, I got an understanding of the organization's structure. Dr Vikram Sarabhai, the chairman of ISRO, doubled up as the director of SSTC, and he had a team of five senior divisional heads—Dr V.R. Gowariker (he later became the director of VSSC:1979–85; and secretary, Department of Science and Technology:1986–91), Dr Gupta (director of VSSC:1985–94), Dr A.E. Muthunayagam (director of Liquid Propulsion Systems Centre:1985–95; secretary, Department of Ocean Development: 1995–2001), Dr Y.J. Rao (chief of Sriharikota Range: 1972–76) and Dr M.K. Mukherjee. This team was responsible for management of day-to-day activities and was chaired by all the members in rotation.

Thumba Equatorial Rocket Launching Station was headed by test director Mr H.G.S. Murthy and two divisional heads, Mr A.P.J. Abdul Kalam (he doesn't need an introduction) and Mr R. Aravamudan (director of Sriharikota High Altitude Range, or SHAR: 1989–94; director of ISRO Satellite Centre, or ISAC: 1994–97). Mr M.R. Kurup was manager of the Rocket Propellant Plant (director of SHAR: 1985–89), while Mr D.Easwara Das was manager of the Rocket Fabrication Facility. Obviously, Dr Sarabhai had created a rare galaxy at Thumba with many stars to its credit.

Hands-on with Inertial Sensors

Dr Gupta, who had joined SSTC as one of the most senior engineers in 1965, was responsible for development of inertial sensors, navigation, guidance and control systems that form the brain of a rocket. He had a brilliant team of nearly 100 members then.

I still admire Dr Gupta's vision and deep insight to organize such a comprehensive set-up in the late 1960s when space technology in India was in its pristine infancy. Dr Gupta, who retired in January 1994 as director of VSSC and member of the Space Commission, currently lives in Thiruvananthapuram. He continues to give his input on the launcher technology and R&D management.

During the first two months in the gyro lab, I spent quite a lot of time in the library apart from working on a few hands-on tasks that Mr Vedachalam and Dr Kulkarni assigned to me.

Little did I know that the initial days of euphoria would soon be over and I would be coupled with a super workaholic boss—Mr Dhas, who had just returned from NASA after a two-month-long training. On the first day after his return, he looked around and found me in the gyro laboratory. He escorted me, like his own child, to his laboratory on the ground floor.

Mr Dhas was heading a fifteen-member team for the development of electromechanical devices as well as pneumatic and hydraulic control power plants for the SLV-3 project. He had specialized in electrical machines at the postgraduate level from Kerala University. An extremely meticulous engineer, he had developed the tiny electric motor operating at 24,000 revolutions per minute (called the Hysteresis Synchronous Motor) that formed the decisive spinning rotor of the rate gyro. This was a critical component in India's first launcher, the SLV-3.

Mr Dhas introduced me to Dr B.N. Suresh and Dr V. Krishnan, the two senior-most members of his team. My camaraderie with these two brilliant and dignified human beings grew during the late 1970s when the three of us returned from higher studies and flocked around Dr Gupta. On 30 November 2007, Dr Suresh handed over the baton of directorship of VSSC to me in the presence of a beaming Dr Gupta; it was a memorable moment. This bond is still going strong.

By the end of 1971, I had completed a brief phase of literature survey of special purpose electromagnetic devices such as the brushless DC motor, servomotors and synchro transducers. My immediate senior, Mahadevan, had plans to leave for higher studies to the US, and he too took special care to groom me so that I could shoulder his responsibilities.

My first major assignment came from Mr Dhas—to design and build a servomotor. The servomotor was ready by February 1972; it performed as expected. We received appreciation from

Dr Gupta and Mr Aravamudan, whom Dr Gupta showed the product the very next day. That tiny instrument helped the closure of my probation period, and more importantly brought me into Dr Gupta's radar. He 'identified' me, and interactions with him became more frequent and stimulating.

By mid-1972, we had a complete team under Mr Dhas, and all of us moved from the cosy air-conditioned ambience of the SSTC main building to the project complex at TERLS. We were housed in one of the sheds with asbestos roofing, with a lot of sun, sea and sand around. I continued to work on the servomotor; the emphasis was on analyses of the servomotor using equivalent circuits and mathematical models to get a grip on the 'how' and 'why' questions. That gave me an insight into the significance of each design parameter on the output characteristics of the end product. This intellectually satisfying exercise evolved further when a mathematician, E.M. George, computerized the design process by 1973; I made myself dispensable as the design of the servomotor could be then done without me. In the process, I got time to attempt something new.

The Greats of ISRO

The director, Dr Sarabhai, would visit SSTC at least once in two months. Amidst his tightly packed schedule, right from descending from the aircraft till boarding the return flight, Dr Sarabhai used to find quality time to visit the laboratories, interact with the young teams and stimulate them. Every department was eager to demonstrate its latest innovation to Dr Sarabhai.

During one such visit to the Control Power Plant laboratory, I had the opportunity to observe Dr Sarabhai at close quarters, making an on-site decision along with Dr Gupta and Mr H.G.S. Murthy to immediately undertake the first sounding rocket flight from Sriharikota in Andhra Pradesh.

On his last visit to Thumba, on 30 December 1971, Dr Sarabhai attended the foundation laying ceremony of the

Thumba railway station, along with Union minister of railways, Kengal Hanumanthaiah. My boss, Mr Dhas, was very happy with his report on the Control Power Plant that was to be presented to Dr Sarabhai the next morning. That did not happen.

The next morning Dr Sarabhai left us for his heavenly abode. The entire space community at Thumba was shattered. I had managed a glance at the mortal remains of that legendary son of this country, whom we still revere as the father of the Indian space programme.

At that moment, I didn't know that four decades later, I would appear in the role of chairman, ISRO, to unveil Dr Sarabhai's statue at ISRO's Space Applications Centre, Ahmedabad in the presence of Mrs Mrinalini Sarabhai and other family members. Two days after my retirement, I was delighted to receive a call from his son Karthikeya Sarabhai to convey Mrs Sarabhai's message, complimenting me for the 'strong and successful leadership' of ISRO for five years. To me, that was a testimony that their 'baby', ISRO, was still in good shape.

The year 1972 was important for our organization. After the untimely demise of Dr Sarabhai, Prof. M.G.K. Menon took over as chairman, ISRO, for a short while. Major organizational restructuring took place—firstly a new Department of Space was carved out of the Department of Atomic Energy. The Indian Space Research Organization, formed in 1969, became the executive body of the Department of Space. And more importantly, the Space Commission was formed. The reins of all these three institutions were handed over to Prof. Dhawan, the then director of the Indian Institute of Science (IISc) in Bengaluru. He had just returned from a sabbatical, in which he was teaching at his alma mater, the California Institute of Technology, arguably the best institution for aerospace engineering.

His requests for establishing Bengaluru as the headquarters for space activities and his continuation at the helm of IISc were accepted by the government. At Thumba, we saw the consolidation of all the units into one entity called the Vikram Sarabhai Space Centre (VSSC), a joint design by Prof. Menon and Prof. Dhawan. They roped in Dr Brahm Prakash, a distinguished scientist from BARC and a renowned materials scientist, as the first director of VSSC. We could experience Prof. Dhawan's dexterity in science administration as objectives became clearer for us at VSSC. The Satellite Launch Vehicle project was approved with Mr Kalam as its chief.

The year of 1974 brought another restructuring at our level. Mr Dhas, along with his team dealing with pneumatics, was to move into the Propulsion Division. Before this move, we needed to accelerate the development of the free gyro that Mr Dhas had initiated. A free gyro is meant to measure the angular displacements of a flying rocket and it needs a built-in directional reference. Hence, the spinning rotor is attached to a double-gimbal frame. Obviously, both in concept and engineering, the free gyro is a bit more intricate compared to the rate gyro. I was a member of this team. A few of us took it up as a challenge and worked on average eighteen hours a day, seven days a week. We received unstinted support from Dr Gupta during this period; he used to drive down to our laboratory from SSTC's main building, late in the evening, just to encourage us. Finally, we demonstrated the working model of the free gyro to Dr Gupta before Mr Dhas moved out on 31 March 1974. That was my first experience of leading a team in the real sense.

This was a turning point for me. Dr Gupta chose me as the engineer-in-charge for the Hysteresis Synchronous Motor (spin motor for rate gyro)—the top priority for the CGI division. The rate gyro was flown on a Centaur rocket as part of the SLV-3 test programme, and I became one of the project scientists for the first flight test of the SLV-3 project. My professional association with Mr G. Madhavan Nair, who later became the sixth chairman of ISRO and my predecessor, started here as he was responsible for the Avionics systems in the SLV-3 project.

The flight went off well. Some of the youngsters like me had the honour of having tea with the chief of the SLV-3 project, Mr Kalam, director Dr Brahm Prakash and chairman Prof. Dhawan. We were thrilled to be personally introduced to Prof. Dhawan, it meant a lot to us. We also got the opportunity to brief the Space Scientific Committee of VSSC, chaired by Dr Brahm Prakash.

Towards IIMB

The desire for higher education was rekindled in me in August 1972 when Mr Dhas, in his inimitable commanding tone, had said, 'Go to Bengaluru, the library in the Indian Institute of Science has some excellent books and journals on rocketry. Get some literature surveys done as we need to develop control power plants for the rockets.' I was very happy for two reasons—first it would be my second visit to Bengaluru after a school trip that happened way back in 1963. And second, I was excited to be in the famed Indian Institute of Science, which had already established itself as a premier educational institute in the country.

Prof. Dhawan was already the chief of both IISc and ISRO. Hence, IISc was like a second home for us. I worked hard; my schedule included ten–twelve hours of reading at the IISc library. The positive vibe was ubiquitous in that campus. The ebullient students and the highly decorated academicians had a great influence on me. In the library, I came across a brilliant engineer, B. Bowonder, who had just finished his doctorate in chemical engineering. I had encouraging interactions with him on the imperative of higher education and possible avenues.

I mulled over the idea for a couple of months. I came to the conclusion that professional advancement in a research and development institution required higher academic qualifications and, more importantly, constant updating of skills and domain knowledge. By that time, arthritis had cast another spell, shackling my hips. So, I had enough time to contemplate in my bed.

After a lot of thinking, I concluded that my engagement in electromagnetics might mould me into a super specialist in a narrow domain among a community of space technologists. I consulted R.B.K. Menon, my good friend with an analytical and philosophic bent of mind. We used to call him '*Asan*' (teacher in Malayalam) and he too endorsed my analysis. The choices for higher studies were then narrowed down to: a post-graduation course either in 'servomechanisms or control system' from an IIT, or 'management studies' from an IIM.

By July 1973, I was promoted to the grade of Engineer-SC with an additional increment for my meritorious performance in the first two years. I became obsessed with the idea of going on a sabbatical for academic purposes. In ISRO, 5 per cent of the employees from each division was entitled to take study leave to pursue higher education in areas of relevance. But it had to go through rigorous scrutiny by a high-level scientific body at the concerned centre; in our case it was the Space Scientific Committee of VSSC, chaired by the director. In those days, exploration of avenues for higher studies was encouraged but the head of the division had to be kept informed.

I was determined to secure an admission at IIT Madras or IIT Bombay for my M.Tech. I could not make it to IIT Madras. IIT Bombay had identified one seat in M.Tech (Control System) for one candidate, to be identified by VSSC. I approached Dr Gupta for his sponsorship, but he thought otherwise and returned my application. He advised me to specialize in electromagnetics. I was disappointed.

Within a month, the admission results of IIM Bangalore were announced. The notification was for their first batch of Postgraduate Diploma in Management, starting in September 1974. In those days IIMB focused on public-sector units, and engineers were preferred candidates. Of course, candidates with work experience enjoyed further preference. Candidates needed to write an essay of 2500 words about the public sector and a shorter one on how the course would help them in the future.

Interestingly, IIMB had put more weight on marks scored in the second and final year of the engineering degree course. This benefited me as my scores were high in both those years. I was still under the shock of my previous application getting rejected. I was so passionate about higher education that I was prepared to resign if nothing worked out.

The admission process, involving a written test, group discussions and interviews, went like a breeze. I got the admission. It was exciting but at the same time I had to tackle two difficult fronts for support and blessings—my beloved parents and Dr Gupta, my divisional head and mentor. I accepted the offer at IIMB and paid the advance fee to seal it.

My father was not too sure if a course in management studies was worth venturing into, letting go of an excellent job at Thumba. I felt I still had time to convince him.

I mustered up the courage and approached Dr Gupta. I felt guilty for keeping him in the dark about this pursuit, even though we met often. I used the rare privilege he had given me and went to his residence to explain. He went through the entire admission file that I had meticulously maintained and read the essays that I had submitted along with my application with a lot of patience.

Dr Gupta did sound a bit disturbed. 'I did not expect this conduct from you, Radhakrishnan. But I am happy that you got through. I will try my best to get you a study leave.' That magnanimous gesture was a lesson for life and a lifeline. That was the first stroke of luck.

I knew that getting a study leave for higher studies in management for an engineer engaged in electromagnetics could probably become an issue during the scrutiny by the Space Scientific Committee at VSSC. They could not take up my case in the next two meetings due to paucity of time. I had no choice but to request IIMB to extend the date of joining.

By then, as a godsend, a letter from the Bureau of Public Enterprises landed up in the office of the director, Dr Brahm

Prakash, soliciting sponsorship for selected candidates from respective employers.

This was the second stroke of luck. Instead of me following up with the administration wing, they started chasing me! Thanks to the magnanimity of Dr Gupta, the positive attitude of Mr M.R. Kurup, the then controller-in-charge, and the swift decision by Dr Brahm Prakash, I was granted a study leave for two years. The leave also ensured continuity of my service and a handsome monthly stipend of Rs 500, but with the execution of a bond.

My father was happy and relieved. I was on my way to Bengaluru for good.

5

MAKING OF A MANAGER

'It's appendicitis, we shouldn't delay the surgery,' said the grave-faced doctor, dressed in a pristine white coat. I was brought to St Philomena's Hospital after I had experienced spells of severe stomach ache the previous night. Prof. Raja Herlekar, the soft-spoken Marathi academician who was also the dean of the postgraduate programme at IIMB, brought me to the hospital with some of my friends. 'Inform his people immediately. I will write to them after the surgery is over,' he directed and asked the doctor, 'When are you planning to operate?'

'Maybe within an hour itself,' the doctor said in a determined voice.

'I hope you didn't tell these guys about the stomach aches you would get in your schooldays. I am sure they have found the ultimate cure this time around,' Arun quipped with a wink the moment I regained consciousness. Arun Balakrishnan, a junior from engineering college, had become one of my closest friends and later room-mate at IIMB. In fact, three more close friends, Balabhaskaran, Kannan and Sankar, were with me in the hospital.

I was discharged from St Philomena's Hospital with eight–nine stitches. At the time of leaving, I thanked Prof. Herlekar for his parental care. My friends had also taken excellent care till my mother and uncle arrived. This was in March 1975.

Pushing Our Limits

The tale of hardships, friendships and cherished memories at IIMB started on 15 September 1974 when I finally enrolled there as a PGDM student. The formal induction programme was over and classes were about to begin. It was a wonderful batch of fifty students with whom I connected almost immediately. Arun, Sankar and I became very close because of our old connections. I shared a three-seater room in the hostel with Balabhaskaran and Kannan for the initial period before becoming Arun's room-mate.

IIMB was taking shape under the great institution builder and management pundit Prof. N.S. Ramaswamy. He had already set up a temporary campus in a rented granite stone building that belonged to the St Joseph's College of Commerce in Shanthi Nagar, Langford Road. This acted as the main building. A few more buildings in the surrounding areas were rented to house the faculty offices and consultancy projects.

A heritage building in the heart of Bengaluru, the temporary campus had an excellent ambience. Even in those days, we had air-conditioned classrooms with special armed chairs for students, and a very eminent pool of faculty members who closely interacted with students.

The pedagogy was notably different from what some of us had gone through in the traditional university system. We had to adjust quickly. We had six terms of three months each, with one of them exclusively for project work.

We, engineers were swiftly drawn towards non-technical yet exciting subjects such as economics, finance, organizational behaviour, operations research, production, operations, projects, marketing, Indian environment and management information systems.

At VSSC, I was used to a tenacious routine of long deliberations and stringent schedules. There were days when I would report at 6 a.m. and return at 10.30 p.m.

But here at IIMB, the academic pressure went far beyond that. By the end of the second term, we had proved to the faculty members that we could remain alert and active even after eighteen–nineteen hours of lectures, group discussions, seminars, projects, take-home assignments and fortnightly tests. We grumbled without effect. The faculty joyfully expressed that the idea was to stretch each of us to the limit and still ensure that we continued to perform in that condition. We actually experienced it, especially when we looked at the quality of the overnight assignments that we had churned out.

The hostel was about three kilometres from the institute, in a couple of rented buildings, at Jayanagar—between the historic Ashoka Pillar and the south-west limits of the iconic Lalbagh Botanical Gardens. The accommodation was of enviable standards. Evening strolls in the Lalbagh gardens on the weekend helped us remain fit and ready for the next week.

Building No. 356, Ashoka Pillar Road, First Block, Jayanagar (currently Lalbagh Nursing Home) is a nostalgic place for me and my friends as we stayed there for over a year. We were a closely knit group, helping each other evolve. Our debating skills were developed in the regular sessions on the terrace of this building. Healthy competition with peers, mostly IITians from the adjacent building, helped us hone our skills further.

Coincidently, four from this debating squadron received the 'distinguished alumni award' from IIMB. Arun (former CMD of Hindustan Petroleum Corporation Ltd) received it in 2008, M.S. Zaheed, (former CMD of HMT Ltd) received it in 2009, I received it in 2010, and Abhishek Mukherjee (founding member of Compaq India in the mid-1990s) received it in 2011.

One of the spin-offs from the hostel life was my initiation into Carnatic vocal music in June 1975. A violinist, Vidwan H.S. Radhakrishna, used to give music lessons to a senior engineer of Indian Telephone Industries (ITI Ltd), living in the adjacent house. Within months, my monotonous voice started settling over the middle octave. It stretched ever so slightly either way, towards the higher and lower notes. My beloved room-mate Arun should be saluted for his tolerance; he lived with my singing sessions and the background sound of an old harmonium. He has reason to feel proud listening to me now.

But thanks to appendicitis, I missed the final examination of the second term. I was cleared for the third term based on my class performance throughout and the special tests conducted after the vacation. I did fairly well in the twenty-four courses required to complete the programme, although I could not get into the first strata. Maybe, the job and financial security had made me a bit indolent to walk that extra mile.

But the scenario quickly changed hue for the individual project that we had to undertake for one full term. I wished to do it at the ISRO headquarters and Dr Gupta directed me to Dr V. Siddhartha, a brilliant and vociferous scientific staff officer to the chairman, ISRO. Prof. Kalyan Singhal, known for his exacting standards, was chosen as the guide from IIMB. The chosen topic was a sensitive one—technology transfer and production of a range of indigenous transducers in a facility set up for export production with foreign know-how.

During the next three months of work at the ISRO headquarters and the brief period of report writing that followed, I spared no efforts to rise to the high standards demanded by the two guides; though, at times, I felt that I was transiting between a lion at IIMB and a tiger at the ISRO headquarters. It was a comprehensive treatise; I secured 'excellent' grades from IIMB, copies sent to ISRO also received commendation. Dr Siddhartha inducted me later, to the Technology Transfer Group that he set up at the headquarters.

An Unusual Decision

We reached the final term of PGDM. All students were asked to attend campus recruitment. I opted out of it. There was a good deal of cajoling, citing the ensuing professional growth, career prospects and remuneration package in public-sector undertakings and government agencies for IIM students.

My well-wishers said that it would be unwise to go back to an R&D institution after being armed with a professional management degree from an IIM. There was also a suggestion that the institute would write to the director of VSSC asking him to give me an assignment that would be commensurate with my management education; I declined all of it politely.

My rationale for this decision was simple: ISRO (through VSSC) had facilitated and supported my higher studies. When I had secured admission on my own to the PGDM programme, three noble men had approved the study leave so that I could pursue higher studies in management. It would have been a travesty of trust if I had left this organization in search of greener pastures. These noble men trusted my judgement on the relevance of management education for ISRO and I could not break that trust. That was one of the best decisions I have taken in my professional life. Time proved it.

I strived hard to avoid the stereotype of the 'typical MBA' and tried to blend into the ISRO family of engineers and scientists. I wanted to demonstrate my management skills only through my work. Some of my well-wishers were apprehensive that the management education may ruin the engineer in me. I ensured them that this would never happen. I remained a student throughout. That is how I managed to switch domains over the decades.

I am often asked the question: How did a management degree help me at ISRO? I feel that a formal degree in management helps one to climb the organizational ladder faster. There are a few naturally gifted individuals who evolve themselves, but I believe the majority of us are not born leaders and have to be trained to become leaders. The degree gave me a better perspective of institutions and issues.

Indeed, the space endeavour is a complex, high-risk, multi-disciplinary, large system that functions in a highly challenging physical environment. But one appreciated the intricate organizational environment (the political, social, economic, financial, environmental, legal and regulatory regimes that constitute it), the human aspirations of its members and its alignment with overall organizational goals, outcomes and their impact, etc. I think in that regard my management education helped me immensely.

IIMB remains very close to my heart. I never miss an opportunity to go there. I am grateful that they have recognized my career and whatever little achievements I have made. Their gestures of inviting me and felicitating me have always given me strength and encouragement.

6

ALLIANCE WITH AVIONICS

VSSC went through a significant structural alteration in early 1976. Dr Gupta was elevated as director of the Avionics Group, and electronic systems were added to his set of responsibilities. Essentially, the Avionics Group comprised four technical divisions: development and production of electronic systems for launchers; inertial sensors and navigation systems; guidance and control systems; and a remnant team of the satellite systems division.

After my return from IIMB, Dr Gupta instituted me as a member of his Progress Monitoring and Industrial Liaison (PMIL) cell. Vinay Auluck, an alumnus of IIMA (1971), was already there. Apart from Auluck, four PGDMs from other IIMs had joined VSSC in 1971, but three of them had resigned. Thomas George, the lone ranger, was doing well in the SLV-3 project. In this situation, I had to tread very carefully in my new avatar.

It was a time of gruelling engagements for all of us. My past acquaintance with many of the senior colleagues helped me get through. We thoroughly enjoyed the change our group was going through.

The Avionics Area Board, under the leadership of Dr Gupta, was looking far ahead on the horizon to give shape to the objectives and priorities of the Avionics Group besides developing VSSC's computational infrastructure and space physics activities (the forerunner of the Space Physics Laboratory established later at Thumba). Back then, the Avionics Area Board (functional since 1974) had the dazzling stars of VSSC, most importantly Mr Kalam and Mr Aravamudan, besides heads of the stakeholding groups and technical divisions. In place of Auluck, who had gone on leave, I had to stand in as secretary of the board, comprehend the crux of the proceedings, and bring out a succinct summary in the end. I had to be a contributing engineer in a domain that was relatively new to me.

Soon, I got into the groove and was later nominated as the secretary of the Avionics Area Board. It was a great experience to work with these mavericks and observe their thought processes.

While I could see the forest, I kept counting the trees too!

Growing with the SLV-3 Project

The Avionics Group had the onerous responsibility of developing and delivering quite a wide range of complex systems for the SLV-3, the first launch vehicle India had developed. Most of the systems had been initiated on the strength of proof of concept, and had to be completed within impossible deadlines. This work culture still brings out the best in ISRO employees—discuss, debate on the technicalities and once finalized impose an impossible deadline. More often than not, the excellent scientists of ISRO invariably deliver. This one brilliant attribute stands out and makes ISRO perform beyond expectations.

The chairman of ISRO used to conduct periodic techno-managerial reviews of projects and technical groups in the 300-seater auditorium at VSSC, which used to be fully packed during these reviews. As the deliverables from the Avionics Group were getting in the way of the SLV-3 schedule, we went through a thorough review in May 1977.

I was allowed to bring in the latest industrial engineering practices in the Avionics Group, besides pursuing my prime role in resource planning. Soon, Dr Gupta gave me the responsibility of managing the delivery schedule for the entire group. The manager in me got excited with this opportunity. We introduced a 'multilevel activity planning and control process', along with a day-long monthly review. D. Narayanamurthi from the SLV-3 project and I worked in tandem to match the priorities and the delivery schedule of the project. Mr Kalam and Dr Gupta were the happiest at this turn of events; the conflicts between SLV-3 and the Avionics Group too vanished. My professional bond with Mr Madhavan Nair, the then project manager of the Avionics systems in the SLV-3 project, blossomed during this period.

Mr P. Ramachandran of the Avionics Group introduced a digital telemetry system (using a pulse code modulation technique) to monitor the health parameters and performance of the SLV-3 launcher during its flight. Matching ground equipment was essential at the ground stations of the Sriharikota launch complex to receive and process the flight data. Mr Ramachandran came up with an indigenous design for the ground equipment too, but his challenge was to produce a dozen of them within a short time; we did not want to go for the easier option of importing them. I was roped in to assist him with production planning methods.

After the launch of the Aryabhata satellite in 1975, the ISRO Satellite Centre in Bengaluru got busy making a satellite for earth observation (Bhaskara-I). ISRO's first communication satellites— the APPLE and the INSAT-1 series—were also on the drawing board. Mr Ramachandran had close professional interactions with both APPLE and INSAT-1 teams. During this time, I learnt about the nuances of satellite communications from him and that came in handy when I moved to the ISRO headquarters a few years later.

Mr Vedachalam's inertial sensors team was developing a rate integrating gyro for which a few components had to be machined out of beryllium to satisfy dimensional stability over a wide temperature range. But beryllium is toxic in nature and difficult to handle; inhalation of beryllium-containing dust can cause a chronic life-threatening allergic disease, berylliosis. The challenge was to set up a precision machining facility where the dust from each machine would get captured at the point of origin, conveyed through a ducting system and then treated and disposed of safely. Ashirvad Geddam, head of precision fabrication in the Avionics Group, was mandated to set up this facility and I was asked to help him. BARC had gained expertise in handling beryllium and they were setting up a Beryllium Extraction Plant at Vashi in Navi Mumbai (earlier Bombay). Hence, ISRO chose to work with BARC and the Beryllium Machining Facility was set up at Vashi. I had the fortune of spending a lot of time with Geddam, from 1978–1979. The result of this teamwork was that I got a good teacher, who taught me a lot about machine tools and metrology.

In the meantime, I felt it was time to work on tangible outputs from my project work at IIMB in the domain of technology transfer and space-industry interface that had been well organized under the ebullient Dr Siddhartha. His Technology Transfer Group at the ISRO headquarters now had its clones in all centres; at VSSC, Auluck was the prime mover for Avionics elements. I assisted him initially and succeeded him in 1980.

My immediate contribution came in three specific cases. First, I assisted S.K. Banerjee, a senior engineer of the Avionics Group, in the transfer of transducer technology to the private firm Encardio-rite, based in Lucknow. Second, N. Sivasubramanian, another senior engineer, and I teamed up to oversee the production of VSSC's rate gyro at Hindustan Aeronautics Limited (HAL), Lucknow, though it did not last long. Third, Kerala State Electronics Development Corporation Ltd (KELTRON), Thiruvananthapuram, was roped in for production of several electronic components. Auluck and I pushed harder in this direction, as this model of partnering with the electronics industry proved fruitful. It was a very satisfying experience of putting theory into practice.

From SLV-3 to ASLV and PSLV

Time flew, as I kept myself busy in several roles; difficult and diverse assignments made me confident of my abilities. This facilitated working closely with several senior engineers beyond the ambit of the Avionics Group. While assisting Dr Gupta in the biannual budget discussions and techno-managerial reviews relevant to the Avionics Group, I got the opportunity to observe Dr Brahm Prakash conducting the Space Centre Council meetings with stalwarts like Dr Gowariker, Dr Gupta, Dr Muthunayagam, Mr Kalam, Mr M.R. Kurup, Mr D. Easwara Das, Mr Aravamudan and Dr S. Ramnath. In November 1979, Dr Brahm Prakash relinquished his seven-year-long distinguished and decisive tenure as director, VSSC. He accepted to continue as a member of the Space Commission.

Dr Gowariker and Dr Gupta, both in their early forties, were the most senior in VSSC, having worked together and traversed as equals since 1966. Eventually, Dr Gowariker was appointed as the director. Dr Gupta reacted in the most professional way, transforming himself into a close, trusted associate and anchor person for the new director. Perhaps his resolve to make India self-reliant in inertial navigation systems was stronger than his personal ambition. He provided unconditional support to the new director to safeguard the interests of the organization. It was a lesson of a lifetime on professionalism and work ethics.

In August 1979, following the unsuccessful experimental flight of SLV-3, VSSC was learning from its failures and preparing for its second flight within a year. Dr Gowariker and the project director, Mr Kalam, would conduct a monthly review in the 300-seater auditorium on technical and schedule-critical issues; the outcomes were monitored closely. I felt honoured when

Mr Kalam chose me as the convenor for one such monthly review, probably the only occasion when someone from outside the project core team had that opportunity.

Around this time, the top management of VSSC was seized with concerns of a likely lull in the activity levels, as the SLV-3 project was in its final phase and new projects were yet to take-off. They were worried about the motivation levels of the scientists and engineers dipping. I conducted a professional study and made a set of suggestions. Dr Gupta endorsed my report and passed it on to Dr Gowariker who was impressed with the report and the author too. From then onwards, I became a member of Dr Gowariker's pool of resource persons at VSSC.

The success of SLV-3 on 18 July 1980 was a historic moment for the country and a turning point for Indian launch vehicle technology. VSSC opened up a new phase into the future with the ideation of the Polar Satellite Launch Vehicle (PSLV) under Dr S. Srinivasan as the study director. Also, he was heading a team of scientists, who were working on the Augmented Satellite Launch Vehicle (ASLV), the next step from SLV-3 and a test bed for new technologies of PSLV. I was inducted as a member of the specialist team, working on 'programme management, manpower and cost' for both these projects.

During this process, I strongly felt that it was the right time to expand the value chain of the Avionics Group by shouldering further responsibilities like system engineering rather than continuing as a 'supplier' of individual packages. Those days, the system engineering activities were handled by the core teams of respective projects.

I expressed my idea to Dr Gupta and he readily welcomed it. A meticulous person, Dr Gupta immediately demanded a concrete action plan. Our case became stronger when Mr Madhavan Nair, the project manager of Avionics in the SLV-3 core team, was moved into the Avionics Group, in 1980, to head the Electronic Systems Division.

Clarity and consensus emerged within the Avionics Group. A team comprising Mr Madhavan Nair, Mr Vedachalam, Dr K. Sudhakara Rao, Mr S.B.R. Shenoy, Dr Suresh and me as the convener was put together to plan for the delivery of Avionics systems for SLV, ASLV and PSLV projects. We took another initiative in the Avionics Group. The new projects would call for three times greater capacity for production of electronic packages and for system integration. Mr Madhavan Nair, Dr Sudhakara Rao, Mr Shenoy and I made a blueprint for a space electronics complex. By any professional standards, it was an excellent project report. We wanted it to come up near Thiruvananthapuram but finally it took shape as the ISRO-BEL complex in Bengaluru in 1983. It was gratifying that our action plan for shouldering system-level responsibilities was accepted by the ISRO headquarters in 1982. The 'Inertial Guidance and Electronics Project' was constituted with Dr Gupta as the chief executive, with both Mr Vedachalam and Mr Madhavan Nair as the project directors, heading inertial guidance and electronic systems respectively. Dr Suresh and Mr Shenoy held these batons a few years later.

These new developments spelt the need for 50 per cent addition to the technical manpower. A high-powered committee comprising Dr Gowariker, Mr T.N. Seshan (the then additional secretary of the Department of Space) and Mr Y.S. Rajan (the scientific secretary of the ISRO headquarters and the eyes, ears and hands of Prof. Dhawan) had been set up to critically review and recommend the additions. The respective heads had to defend their demand for the proposed manpower.

Dr Gupta had to step out that day, and hence he nominated me to defend our case. I made a crisp presentation and defended our case by answering all questions put forth by the high-powered committee. Both Mr Seshan and Mr Rajan were at their incisive best that afternoon. Later, I got feedback that the committee found our demand to be relevant, and they were happy with the way I presented it. It was gratifying.

The Turning Point

My residency period as Engineer-SD was coming to an end. The going was great at the office with challenging assignments and a nurturing boss. I came to be known as a live wire. On a personal front too, life at Thiruvananthapuram was easy and comfortable. My brother, Sivadasan, and I had bought two adjoining pieces of land, ready to build independent houses.

Thiruvananthapuram had a rich legacy in Carnatic music and Kathakali, and I did not want to let go of my love for music and rhythm. I had been taking serious lessons in Carnatic vocal music and watched Kathakali performances whenever I got free time. I had had a few stage performances in both vocal music and dance. I had done my best in both. In fact, I had started looking like an artiste, sporting a beard and a sling bag.

This was probably an ideal setting for complacency. But there was a constant inner call to rupture this comfort zone and wake up before professional stagnation sank in.

Destiny has its own curious ways. Dr K.S. Prabhu, one of the four scientific staff officers of the chairman, Prof. Dhawan, in charge of the budget management for the entire ISRO, decided to go on a sabbatical.

On a sultry forenoon in March 1981, I got a telephone call from none other than the young and vibrant scientific secretary Rajan. He was brief and direct, 'Prof. Dhawan has zeroed in on your name for a position at the ISRO headquarters. Do you have any problems with it?' I was a bit taken aback with that approach. I tried to be as convincing as possible, 'I have no problem in moving to Bengaluru if Dr Gupta concurs.' Mr Rajan responded, 'Leave it to me. I will handle it.'

This encouraged me further and I added, 'I would prefer to move only after my promotion interview, expected to take place by the end of this month.' He accepted it. I hinted to Dr Gupta about the development. He said, 'Let's see.'

It seems Prof. Dhawan had written a nice letter to Dr Gowariker, asking for me to be immediately moved to Bengaluru. A new office of Budget and Economic Analysis (BEA) had been set up with P. Sudarsan as its director and I was appointed as its manager.

In those days, it was rare to promote an Engineer-SD grade officer or even a higher-level officer to the position of manager. A cabin at the headquarters was kept ready for me. In every sense, it was an offer that could not be refused.

The parting advice to me from Dr Gupta was, 'I know that you hold a strong viewpoint on some of the current issues, rightly from the perspective of the VSSC. But if you come across more information at the headquarters, feel free to take on an objective view even if you have to deviate from the current one. I will be happy if you do that.'

Such advice could emanate only from an evolved soul. I engraved those words in my mind as I prepared to leave for Bengaluru.

7

BLOOMING AT BENGALURU

'Join the conspiracy!' Prof. Dhawan exclaimed in his inimitable style, as Mr Y.S. Rajan introduced the latest entrant to the ISRO headquarters. Prof. Dhawan, still the director of the Indian Institute of Science, was at the helm of India's space programme; he donned three hats i.e. secretary of the Department of Space (DOS), chairman of the Space Commission and chairman of ISRO.

A few days earlier, SLV-3 had launched the Rohini Satellite RSD-1, the first experimental communication satellite that APPLE was slated to lift from French Guiana within the next ten days. Amidst these crucial events, Prof. Dhawan was planning to organize a meeting to discuss the configuration of the PSLV, the flagship launch system of ISRO, with the project director Dr Srinivasan, programme director Mr Kalam and scientific secretary Rajan. I didn't know then that this was the beginning of a lifelong bond with Prof. Dhawan. I also didn't know that I would be his last student and would later occupy the chair that he decorated with distinction for over a decade.

In the second week of June 1981, I landed at the Bangalore City railway station with a suitcase, two cartons of books, and a lot of excitement. The ISRO headquarters were just four kilometres from there, comprising nine floors and housed in the F-block of

47

Cauvery Bhavan office complex on Kempegowda Road. I had been there for my PGDM project assignment in 1976, so the turf was a familiar one.

I had to report to Sudarsan, director of the newly set-up Budget and Economic Analysis. Sudarsan, a brilliant chemical engineer and a PGDM from IIMA, had been at the headquarters for the past five years, coordinating programmes for development of materials and propellants.

Sudarsan introduced me to the main actors at the headquarters. The illustrious Mr Seshan and his junior Mr K.A. Varadan from the Indian Administrative Service jointly held the flag of Department of Space. Dr Siddhartha, my project guide of 1976, headed the technology transfer and industry interface function. He wanted me to assist him too. Three veterans and my well-wishers from Thumba, Mr Kalam, Mr M.R. Kurup and Mr Aravamudan, were part-time functionaries at the headquarters, managing launch vehicle programmes, safety and reliability functions. Sreenivasa Shetty and P.N. Jayaraman, two other members at the headquarters, were familiar faces from my days at Thumba. Jai Singh managed the satellite communication programmes and the INSAT system. About fifteen young engineers, with specific domain expertise, assisted these seniors.

There was palpable warmth from everyone I came across in that compact and cohesive monolithic headquarters of the Space Commission, DOS and ISRO. It was a good start to a sixteen-year-long innings.

Budget and Economic Analysis: Not Just Arithmetic

By the 1980s, ISRO was in transition—it was past the experimental phase and was developing satellites which were going to be a vital aspect of the national infrastructure. The INSAT-1 satellites for communications, broadcasting and meteorological services were being built abroad and an INSAT Coordination Committee was overseeing this multi-departmental endeavour. The Indian Remote Sensing (IRS) satellite, to be built by ISRO, and the PSLV, aimed at self-reliance for launching IRS satellites, were on the drawing board. The National Natural Resources Management System (NNRMS), using IRS, was also being developed.

Development of the PSLV was a major leap forward for ISRO, both in terms of the launch capacity (the mass of satellite being launched) and the accuracy of the orbiting satellite. The SLV-3 could, at best, orbit a satellite with a mass of forty kilograms, whereas IRS satellites of 600 kilograms (and even more) had to be placed precisely in the specified orbits. The ASLV had been taking shape as an intermediate step with launch capability for 150-kilogram satellites. Besides, it provided a test bed for some of the new technology elements (e.g. strap-on staging with the booster rocket and closed loop control system) for the PSLV.

The ingenuity of Prof. Dhawan was evident in the design of the headquarters with diligent mechanisms, processes and people to handle the three-in-one role assigned by him. Every functionary had clarity on the role, boundary conditions, responsibilities and accountability within.

It was customary for members of the headquarters to generate and circulate analytical position papers on the critical issues and future perspectives in their domains. The ensuing frank discussions helped to bring out the collective wisdom of all actors concerned from ISRO and DOS before major decisions were taken.

I soon found a place for myself at the headquarters, thanks to my background at VSSC. I highlighted priority clashes in the multi-project environment around launch vehicles and made a few concrete suggestions. I was glad when I came to know that Jai Singh held me in high esteem. 'He is headquarters material,' he had said even as Sudarsan had kind words, 'The loss of VSSC is headquarters' gain.'

Rajan Sahib (I chose to address him so from here onwards) and Sudarsan primed me for the imminent plunge into the annual budgeting cycle, spanning over July–December every year.

A recent training in Bengaluru had introduced me to public finance in India. But the process followed in DOS and ISRO was more comprehensive and unique in many ways. It focused on the linkages between programmes, schedules and budget as well as the associated techno-managerial decisions within the framework of a long-term perspective plan that was already in place. My role as an experienced technical manager was indeed a demanding one, and both these gentlemen became my co-guides in the days to come.

The budget cycle of 1981 kicked off in the first week of July, and we had to make a realistic assessment of funds for revised estimates for ISRO's programmes on hand and come up with the programmatic targets and budget estimates for 1982–83. This exercise required BEA to work in tandem with all programme offices of ISRO headquarters and Mr Seshan's team from DOS. I became Sudarsan's understudy, to draw up programmatic and financial guidelines, essentially the framework for a month-long exercise to be followed at the centres. This was followed by a joint review at the centre by the centre director and the headquarters team led by Mr Seshan. It was my task to keep records of the long discussions. These reviews were intense with analysis and logic and at times stormy, but by the end consensus emerged on inclusion of a majority of proposals and disagreement on several to be escalated to the level of the ISRO council.

The ISRO council, with directors of ISRO centres, additional secretaries and joint secretaries, was a cleverly crafted mechanism to deliberate regularly and religiously over all programmatic and techno-managerial issues. This body ensured congruent decision-making at both ISRO and DOS secretariats, within the ambit of policies laid down by the Space Commission.

A day-long session at Bengaluru with stakeholders and nearly 100 senior professionals from different ISRO centres was an essential prelude for the ISRO council to firm up its decisions. BEA administered the process and the baton was passed on to the DOS secretariat for the next steps in coordination with the Planning Commission, Space Commission and Finance ministry.

At the close of the first budget season of five months, I got a panoramic view of the entire gamut of India's space programme, its prime actors, interfaces and issues. Also, I realized that I had been accepted across ISRO, DOS and its headquarters.

I managed to live for a couple of months in the guest accommodation of ISRO's Auxiliary Propulsion System Unit (currently LPSC) located at Jeevan Bhima Nagar, thanks to its director, Dr Muthunayagam. Later, Mr C.A. Balasubramanian, the chief controller of accounts of DOS, another recent entrant at the headquarters, graciously invited me to share his rented accommodation at Kumara Park. Mr Balasubramanian and his wife, Prof. Saraswathi, looked after me like their son. The house owner, Mr K.M. Murarappa, a retired chief engineer, who stayed on the ground floor of the house, epitomized dignity and honesty. That was my home away from home till I settled down after marriage in 1983. It was just a five-minute walk from my music guru R.K. Srikantan's residence. This gave me ample time to take lessons, sit through his practice sessions, escort him during performances and handle his correspondence, all in the *gurukula* style.

During the first half of 1982, I got a break at the headquarters. A major gap between the Five-Year Plan allocation for 1980–85 and the demands of emerging projects had made techno-economic analysis an imperative prerequisite. My past experience in VSSC helped me suggest a restructuring of the work packages of the PSLV mega project and thereby gain a possible cost reduction by 10–15 per cent. This could be used to fund two more projects. Initially, when this analysis was presented, Prof. Dhawan had commented, 'You have done your homework well.' He then decided to get it reviewed by Dr Brahm Prakash, a member of the Space Commission. I was then chosen as the member-secretary of the team set up to assist Dr Brahm Prakash. Both Kalam Sir and Dr Gupta were members of this team besides a few other colleagues from the headquarters. It was a unique recognition and a rare experience. Most of our arguments were accepted by the project teams and the management.

For injecting the IRS satellite (or any satellite for that matter) into the specified orbit with allowable dispersion in six orbital parameters, the PSLV had to first navigate itself precisely. This was done by an Inertial Navigation System (INS), which is used in aircraft too. Self-reliance in this strategic technology is most crucial for Indian launch vehicles in the future. The IRS and INSAT satellites too needed an inertial reference system to complement the star-based navigation systems. The Avionics Group had the mandate to develop it. Many brilliant technologists came up with options for the type of gyro to be developed and the schemes for INS. The choice was difficult, as the process was getting protracted. Soon, it snowballed into a contentious issue. It was pushed to the headquarters for a decision at the chairman's level.

By then, I had completed a year at the headquarters, with a reasonable level of detachment from the Avionics Group to be able to look at one of the complex technology decisions in a broader perspective. I wrote a comprehensive position paper which Rajan Sahib endorsed. Prof. Dhawan sat with us for a few sessions, like a doctoral guide, to address the technical merits of the options, intricate technology issues and the human dimension.

The decision, articulated as a fairly long office memorandum in mid-1982, was welcomed by all stakeholders and it has stood the test of time, even after three decades. I was appointed as the representative of the ISRO headquarters in the Project Management Board of Inertial Guidance and the Electronics Project Board, which I shouldered till I became the executive chief of the five regional remote sensing service centres.

Changing Gears

Rajan Sahib started engaging me for sensitive management system analysis as well as for assisting him on matters related to the

ISRO council and the Space Commission. An excellent engineer, manager and administrator, he became my mentor. The intense interactions with him in the office and during tours, helped me to imbibe some of his traits and talents. I felt that I was turning into one of his trusted lieutenants. Some commented that a 'junior Rajan' was in the making. It is true that, back then, Rajan Sahib and I looked alike.

I discovered another mentor in Mr Seshan as he began to rope me in for analysis of procurements and finance functions. It was an experience to observe Mr Seshan's style and to strive to come up to his exacting standards.

The masala dosa and basundi from the adjoining Woodlands Hotel and a stroll around Cauvery Bhavan kept my energy levels and spirits high enough for long evening sessions at the headquarters.

Interactions with Prof. Dhawan became more frequent. I felt an excellent chemistry between us. That was a turning point. I started accompanying him for meetings in Delhi and Thiruvananthapuram. My flair for graphics was called upon during arduous preparation for Prof. Dhawan's meetings and talks. In the process, I picked up conceptual comprehension.

When Prime Minister Indira Gandhi visited Sriharikota in April 1983, to witness the launch of SLV-3, he had me by his side (along with Rajan Sahib and P.N. Jayaraman). He encouraged me to brief Indira Gandhi for five minutes when the rocket took off (I was thirty-three then). Later, Rajan Sahib commented that the PM was staring at my *rudraksha mala* (prayer beads), which I had been wearing since a couple of years.

By July 1983, Mini became my life partner. Both of us hailed from Irinjalakuda and our families had been acquainted since three generations. Within a week of my marriage, I was selected for a three-week training visit to the major firms of French space industry, engaged in the development of the Ariane launch vehicle, communication satellites and other components. Many at home saw this as a good beginning.

Mini was then employed at the State Bank of Travancore in Irinjalakuda, just a five-minute walk from her house. She readily agreed to move to Bengaluru, but the process of interstate transfer was a major hurdle. Mr Seshan dropped into my cabin and offered to help. I was deeply touched by this gesture. Mini was able to join me at Bengaluru within four months, and we settled in a two-bedroom house in south Bengaluru.

The formulation of the Seventh Five-Year Plan commenced in August 1983 and Sudarsan was put in charge of steering it. I was his understudy in this exercise. We identified a host of resource persons, mostly in their mid-career, to stimulate different perspectives to enhance the 1980–90 decade profile.

When the plan document was cleared by the ISRO council, Mr Seshan threw a challenge at me. He said, 'This document is very good for use within ISRO but it will not sell in Delhi. We need to bring out the national importance for taking up such a massive programme. There is only one person who can write that way. He is sitting in that corner room (he was referring to Prof. Dhawan). Let's help him. Invoke goddess Saraswati and start writing, preferably with a steel pen if you can get hold of one. Do not give dictation.' I did so in letter and spirit. The total transformation of the document was visible; there was a huge change in my style of writing.

In the beginning of 1984, on three occasions of Rajan Sahib's foreign deputation, I was called upon to stand in as the scientific secretary. There were many raised eyebrows as several of my seniors at the headquarters could have also been chosen for it. I got the feedback that Prof. Dhawan was happy with my performance.

Soon after this, in March 1984, Prof. Dhawan called me to his room and said, 'We don't know how to make use of you! Let's discuss this.' He bracketed me with two other IIMA alumnus, considered as high performers at ISRO, and further appreciated that my technological insights were superior to them. That was the beginning of a long and relaxed one-to-one session on my career planning that he undertook earnestly and seriously.

He counselled me like a parent. He summarized his observations of me succinctly and listed out a few options that I could take up. However, he insisted that I should continue with the budget activities for at least one more year.

The option of appointing me as the assistant scientific secretary was raised and ruled out as we felt that executive responsibility was the pathway to success. Two new opportunities opened up for executive responsibilities.

The first opportunity was a role at the Spacecraft Tracking and Control Centre that ISRO had planned to set up in Bengaluru. The second opportunity would come up within a few months, as DOS was called upon to set up a chain of Regional Remote Sensing Service Centres (RRSSCs), under the ambit of NNRMS, co-sponsored by DOS and a few other central ministries. Rajan Sahib was spearheading NNRMS as a platform for application of data from remote sensing satellites across the entire country. I opted for the latter option and waited for my turn.

DOS was chosen to set up the RRSSCs on behalf of the co-sponsors. Prof. Dhawan made up his mind to give me additional executive responsibilities as the project manager. By then, he was preparing to pass on the baton to Prof. Rao. He did not wish to impose that decision on his successor by issuing it a month before demitting office. Instead, he presented it to his successor, Prof. Rao, through the scientific secretary, Rajan, and left it to the successor to decide on its implementation. I was gladly accepted by Prof. Rao for that position. That became chairman Prof. Rao's first office order, issued on 9 October 1984.

This was a great lesson on how an outgoing chief should conduct business.

8

THE CREST AND CROWNS

ISRO was ready for a transition of its leadership. That was in 1984, after twelve years of meticulous planning and organizational structuring under the charismatic Prof. Dhawan. Prof. Rao's ascent to the coveted position was announced in the first week of September 1984.

A well-crafted and seamless transition was in place at ISRO's headquarters. We observed and appreciated as Prof. Dhawan and Prof. Rao started taking joint briefing sessions on a daily basis. Prof. Rao's credibility as a scientist and leadership qualities were well recognized by all.

We had sensed this change when Prof. Rao was named as a member of the Space Commission in 1982 itself. There were one or two voices of dissent from contemporaries, but these did not gather much support. The appointment of Prof. Rao in this new role was widely acclaimed and accepted.

Prof. Dhawan opted to relinquish his new role as the senior adviser at the end of the first year. However, he was persuaded to continue in the Space Commission till 2001.

The assassination of the then Prime Minister Indira Gandhi (31 October 1984) had its repercussions on the space programme since she was the minister-in-charge of the programme. Prof. Rao navigated through this well. In fact, it was quite an experience observing him later, as he waded through changes in the national leadership and associated changes in the Space Commission. All these happened during the three launch failures he encountered.

Those days, ISRO's headquarters was nothing less than a star-studded galaxy. Stalwarts such as Mr Seshan (additional secretary of DOS), Mr Viswanathan (joint secretary of DOS), Rajan Sahib (scientific secretary), Jai Singh (programme director of INSAT), Srinivasa Setti (assistant scientific secretary and director of the Launch Vehicle Programmes) and Sudarsan (chairman of the Technology Transfer Group and director of BEA) were there to support the new chairman.

My professional interactions with Prof. Rao in the past, especially during an intense review of the INSAT-2 project report, helped me adjust better in the new regime.

Into Higher Planes

The annual budgeting cycle was then at its peak. We had our apprehensions regarding the dynamics of the situation that would impact the programmatic re-prioritization and the financial decision-making process. Prof. Rao deftly handled the day-long sessions of the ISRO council, held within a fortnight of his shift to the corner room.

During this process, the seasoned administrator, Mr Seshan advised me, 'Raise your points with a smile. Keep smiling even if someone reacts; finally everyone will accept the reality.' We met our stringent targets at the end of the day.

Zero-based Budgeting (ZBB) was the buzzword in the 1970s, coined by Peter Pyhrr, and famously used by President Jimmy Carter when he was the governor of Georgia in 1973. We had been using this concept in the budgeting process. In essence, the concept implies, 'start from a zero base and analyse every function within an organization for its needs and costs during each budgeting period'.

The Government of India wished to adapt this for other science and technology departments as well, and constituted a committee of financial advisers in 1984. Mr Seshan was its chairman and he roped me in to draft the background paper. I put my heart and soul into it.

Mr Seshan endorsed my paper and inducted me into the National Committee of Financial Advisers. It was an honour to serve in that esteemed committee. Even after he left ISRO and moved to Delhi (in February 1985), I continued to have several interactions with him.

In 1985, the top management of ISRO had its plate full as the government had approved the INSAT-2 project. We observed the spirit of an aggressive project director in Prof. Rao, as he spread his wings beyond his pet satellite projects. He placed equal weight on development of space application projects for national missions, or steering crucial development phases of ASLV and PSLV, or coaxing Indian industries to produce chemicals and other raw materials.

During that time, Prime Minister Rajiv Gandhi had just set up a ministry for programme implementation, under which there was monthly monitoring of key programmes of every ministry and department. I was the focal point in DOS. Our concise yet comprehensive monthly report with a two-page graphic was much appreciated in Delhi.

As a matter of fact, Mrs Sarla Grewal, secretary to the PM, arranged a meeting in December 1985, with all Science and Technology (S&T) secretaries for adapting the DOS methodology on reporting and programme management. That was my first entry into the prime minister's office (PMO). That meeting facilitated five special dispensations to the S&T departments on the lines of DOS, based on a policy paper by Mr Seshan, then the environment secretary in Delhi.

I vividly remember that Friday evening. Mr Seshan had asked me to draft the paper before my return to Bengaluru the next afternoon. The next week, I called on him at his office as I usually did during my trips to Delhi. He took out a paper from his drawer and said, 'Read it. I sent your draft to the PM as such, except that the conclusions have been brought out in the covering note. The PM has approved it in principle.'

That was a magnificent gesture of acknowledgement.

During that time, a cost-efficient PSLV programme was on ISRO's top-priority list. Dr Brahm Prakash had done the first round of pruning of several programme elements embedded under PSLV's cost in 1982 itself. As the development of PSLV progressed, several issues on inescapable costs surfaced. Prof. Rao chose a team led by Mr N. Pant (fondly called Pant Sahib) to keep a check on this.

Pant Sahib was known for his uncanny knack of raising fundamental doubts with childlike enthusiasm. Dr Gupta, Mr Aravamudan and Rajan Sahib were the members. I was chosen as the member-secretary of this committee. The Space Commission took up this agenda in February 1987, and both Dr Srinivasan and I were called in to defend our case. Dr Srinivasan briefed on the technical achievements of the PSLV and I presented the cost escalation.

On several occasions, I was amazed at the trust Prof. Rao had in me even when the subject matter was not in my domain.

By 1985, the prudence of consolidating liquid propulsion development under Dr Muthunayagam became obvious to ISRO's management. Prof. Rao identified me as the headquarters' representative in the high-level committee chaired by Dr Gupta that had been set up to iron out a variety of perspectives on this sensitive issue.

As Dr Gowariker, director of VSSC, was moving out on a sabbatical by the end of 1985 and Dr Gupta was taking over as director of VSSC, I was called in to provide management systems analysis support for a major restructuring, bringing in a new layer of deputy directors in the top management of VSSC. It had been the largest centre of ISRO and the time was considered ripe for this decentralization. Each of these deputy directors was in charge of an organizational unit named 'Entity', comprising homogeneous groups that were to be managed by a 'group director'. This management structure stood the test for over three decades and was implemented in the rest of ISRO later.

I was promoted to the post of Scientist/Engineer 'SF' on 1 January 1986, within the minimum residency period. More importantly, after my promotion interview, the committee invited me to join them for dinner, presumably as a gesture of their satisfaction.

Dr Gupta, the then acting-director of VSSC, was a member of the committee. Later, he called me to the director's room and said, 'Radhakrishnan, I am proud of you. Both chairmen, Prof. Dhawan, and Prof. Rao, are very happy with your performance at the headquarters. I just wanted to congratulate you.'

This was a proud moment for me. Dr Gupta's appreciation meant much more than the promotion itself! Dr Gupta, like a true guru, continued to guide and advise me in the later years, even during my tenure as the chairman of ISRO!

In June 1986, I got the opportunity to accompany Prof. Rao, along with Rajan Sahib and Jai Singh, for a one-hour-long presentation to Prime Minister Rajiv Gandhi and his Finance Minister V.P. Singh in the South Block conference hall. The occasion was to brief the PM on the imperative for a substantial hike (of about 75 per cent) in the allocation for DOS in the Seventh Five-Year Plan (1985–90) to pursue development of INSAT-2, IRS series, PSLV and GSLV. I was seated just behind Prof. Rao and answered two questions raised by the PM. After that, I became one of the three emissaries of Prof. Rao from the ISRO headquarters for relevant briefings at the PMO.

Decoding Remote Sensing

My executive task of setting up RRSSCs at Bengaluru, Dehradun, Jodhpur, Kharagpur and Nagpur had gained momentum in 1985.

This was the second instance when I was changing my domain, and I approached it with great enthusiasm and interest.

In simple terms, remote sensing was all about obtaining information about an object, area or phenomenon through analysis of data acquired remotely. In our case, the data collection was being done through space-borne platforms like satellites and sparingly by aircrafts. My task of setting up the RRSSCs under the ambit of NNRMS, opened up the potential of established satellite systems like the Landsat of the US, SPOT of France and our emerging IRS, as well as to the vastness of its utilization for monitoring natural resources.

This could eventually reduce dependence on treacherous and time-consuming ground surveys. The NNRMS was evolving as a nationally orchestrated system to ensure proper usage of satellite data by Central and state government agencies for management of agriculture, soil, land use, water resources, forestry, minerals, marine resources and environment.

Thus, I was introduced to ISRO's bread and butter—the 'Space Applications' programme, encompassing diverse domains like remote sensing, meteorology and communications programmes at that point of time.

Data (or images) collected for each of these natural resources by remote satellites required interpretation to extract information relevant to managers or decision makers. The 'visual interpretation' of images has been in practice for a long time in India too. Visual interpretation makes use of the ability of the human mind to qualitatively evaluate the spatial patterns in an image, based on shape, size, pattern, tone, texture, shadows, etc. The human eye has inherent limitations to discern all tonal variations of such images. Even small tonal variations could contain immense amounts of information which cannot be ignored. That prompted the need for computer-based classification of the tonal variations or the spectral patterns. Thus, the concept of digital image processing came into existence. Later, it imbibed the essence of the human mind's ability for contextual classification; artificial intelligence and neural networks also evolved. After three decades of continuous evolution, digital image processing now forms the backbone of satellite-based remote sensing.

This, in turn, gave wings to a more sophisticated tool called the Geographic Information System or GIS. Prima facie, GIS is a versatile computer-based tool to synthesize location or spatial data (maps) with all related attributes to assimilate relevant information to support decision-making. The advent of database management systems, navigational satellites and mathematical analysis tools contributed to the versatility of GIS and enhanced its user base.

In the light of such transformational developments, the National Planning Committee of NNRMS realized the imperative to set up regional hubs (i.e. RRSSCs) in the country to develop digital image processing and GIS. By 1985, DOS and three central departments sponsored RRSSCs at Bengaluru, Dehradun, Jodhpur, Kharagpur and Nagpur. I had the prestigious task of delivering these despite several initial hiccups with some of the host institutions. The imminent task was to build a core team with domain experts from other institutions.

Mr K. Krishnanunni, a progressive geologist from the Geological Survey of India (GSI), joined on deputation, as the director of RRSSC and my immediate boss. From the National Remote Sensing Agency (NRSA), S.K. Bhan, a geologist, and Geeta Varadan, an electronics engineer, joined us. Two seasoned scientists from agriculture and soil domains, Dr R.L. Karale and Dr R.P. Dhir, Prof. S. Sengupta, a brilliant geophysicist from IIT Kharagpur, and several experienced geologists from GSI were inducted on deputation.

Nearly 100 young scientists and engineers were inducted. All of us went through a phase of collective learning, training and team building at Bengaluru, before the teams were deployed at four other locations.

There were many challenges. Procurement of nine VAX-11/780 computer systems attracted export restrictions. Youngsters had to be groomed as system managers at a time when only seniors were allowed to handle computers! They turned out to be brilliant— one of them, P.G. Diwakar, is currently the scientific secretary of ISRO. I learnt fundamentals of digital image processing from my young colleague Diwakar. He was an admirable all-rounder at work and even on the stage.

The first phase of our mission was successful as RRSSC processed the first imagery from IRS-1A (launched on 17 March 1988) and comparative studies were made with imagery from the Landsat series of the USA for different application themes. Soon, RRSSCs made use of inroads into digital image processing for (a) generating fortnightly status updates of vegetation for the entire country (b) delineating areas where major crops are grown, thanks to our first major stakeholders from the Space Applications Centre (SAC) and NRSA.

NRSA was the national hub for receiving remote sensing data from IRS-1A and foreign satellites. Another significant step forward at NRSA was to enhance capacity for satellite data processing. Initially, I was inducted as an observer on behalf of RRSSC when a team led by Dr George Joseph of SAC prepared its blueprint in 1986. However, my role was later enhanced as the headquarters' representative in the Project Management Board that oversaw the implementation. That turned out to be a phase of rich learning, working closely with the doyens of that domain—Prof. B.L. Deekshatulu and his colleagues. This came in handy later for my roles at the NRSA.

ISRO's Long-term Planning

By August 1987, Prof. Rao elevated me from manager to director of the Budget and Economic Analysis. It was an exalted position as I was only thirty-eight then. With higher responsibility came more expectations. My designation at RRSSC also changed from project manager to project director. I sent a handwritten note of gratitude to Prof. Rao through Rajan Sahib. Prof. Rao wrote on it, 'A very good person.' That note remains one of my most precious possessions till date.

It has been customary to formulate decadal profiles for space programmes, consistent with the long gestation period for space projects, from concept to fruition and operational life cycle. Prof. Rao set up a compact team with Rajan Sahib as the convener and Dr K. Kasturirangan, Mr R.M. Vasagam, Jai Singh (as members) and me (as member-secretary) to steer the 1990–2000 decade profile.

It took nearly a year for this process as we scanned the global scenario, assessed national imperatives and scientific aspirations, industry's preparedness and technology progression. With the stakeholders on board, the process was analytical, iterative and consultative. And finally, a robust profile was created, with a pragmatic strategy for satellites and their application for communications, broadcasting, remote sensing, meteorology, space sciences, and a self-reliant launch capability.

The indicative outlay for the decade 1990–2000 was Rs 10,000 crore (USD 7.7 billion). The investment being substantial, cost-effectiveness, cost-benefit and economic analysis became essential parts of the exercise. The Space Commission endorsed this profile after a long session of presentations and discussions held at the ISRO Satellite Centre, Bengaluru.

It was then required to garner endorsements from the government and political leaders for the broad direction and elements of the decade profile. Prof. Rao and his team made a presentation in April 1987 to the deputy chairman Dr Manmohan Singh and members of the Planning Commission.

At the end, Mr Nitin Desai, special secretary in the Planning Commission, was requested to examine various aspects related to investments, alternatives, etc. First, he examined the utility and effectiveness of space-based solutions in communications, remote sensing and meteorology vis-à-vis alternate options (e.g. optical fibre communication). He came up with three conclusions: (a) satellite communication still makes sense for remote areas, distributed communication needs and as a source of diversity; (b) the value of remote sensing applications will increase when certain areas, which are experimental at present, become operational and as the capabilities in user organizations are enhanced; (c) as for meteorology, the case was always strong and remains so.

Then, the second aspect of space technology was addressed. Questions pertaining to 'make or buy' were discussed, including hiring of satellite capacity from foreign operators and procurement of both satellites and launch systems from abroad. Finally, Mr Nitin Desai concluded that domestic development of satellites and launch capability also seem to be cost-effective and, in some cases, a strategic necessity. Also, he highlighted the export potential for India in space technology and that such commercial opportunities will make the Indian space engineering industry a more viable proposition.

Today, some of these might be obvious to most, but in the mid-1980s it was a challenge to sell this idea to the decision makers in the government and polity, let alone the public at large.

After this comprehensive exercise with all stakeholders, in November 1988, Mr Nitin Desai (the chief economic adviser by then), endorsed our proposal. But he took us through a rigorous review.

It was an experience for me to work closely with this brilliant economist and soft-spoken human being for six months.

Up Close with Failures

One of the greatest learnings of my professional life came through my association with ASLV as the headquarters' representative in its Project Management Board from 1986. It was a tough time for the organization after the failures of the first two flights of ASLV in 1987 and 1988. I keenly observed and absorbed the composure and mettle of Prof. Rao at close quarters as he went through the agonizing moments of those two failures and the aftermath.

Prime Minister Rajiv Gandhi was present to witness the first launch of ASLV on 24 March 1987 along with the governor and chief minister of Andhra Pradesh, minister of state K.R. Narayanan and twenty parliamentarians. Mr Seshan, who was doubling up as secretary (internal security), was present too. Rajan Sahib and I were with the VVIPs on the terrace of the first-generation control centre (the majestic edifice that we see now is of the third generation).

I vividly remember the launch. The countdown progressed smoothly. The ASLV rocket majestically took off as the two strap-on stages ignited at 1209 hours and completed their job within a minute. There was tremendous jubilation and compliments galore. The worst was in the offing as the first stage of the rocket failed to ignite! Amidst this jubilation, the telephone on the terrace rang and Rajan Sahib got a message from the range director, Jayamani. Rajan Sahib murmured to me in Tamil 'crashed'; both of us kept a straight yet brave face. Soon, the prime minister was escorted to the Mission Control Room. Mr M.R. Kurup, the then director of Satish Dhawan Space Centre (SHAR) and chairman of the Launch Authorization Board, was upfront in his approach. 'Sir, we didn't do our job. Mission failed,' he said. Later it became clear that the first stage did not ignite and the rocket broke off. The safe arm device introduced for range safety in the first stage became the suspect.

The second flight took place on 13 July 1988; this time the safe arm had been eliminated. I was in the control room, behind Mr Aravamudan, with orders to convey the progress of the flight to radio commentators. This time the rocket's first stage did ignite but still the flight failed! We were shattered but immediately sprang into action. Prof. Rao along with the centre directors, Pant Sahib, Dr Gupta and Mr M.R. Kurup, sat with a group of fifty–sixty experts till the wee hours of the next day to study the problem— the how and why of it and the way forward. Kalam Sir, the then chief of Defence Research and Development Organization, also joined the team.

Throughout, Prof. Rao put on a brave face, absorbing the brunt of the situation. He maintained his cool and consoled his shattered colleagues. But, he kept smoking, almost non-stop, probably in an attempt to let off steam. On one occasion, just outside the review hall, I whispered to him, unable to hold back, 'Sir, you are smoking too much.' He quipped, 'What else can I do now?' He threw his cigarette and went back to the review hall. I heaved a sigh of great relief.

I had always heard that history repeats itself, but I experienced it in my own life. Those great moments of learning that I gained by observing Prof. Rao from close quarters during such crises came to my rescue when I went through two consecutive failures of the GSLV during the first year of my chairmanship.

9

ACCOLADES FOR ASSIDUITY

The first half of 1988 saw major changes in the headquarters. It started with the exit of Sudarsan who moved out to set up the 'Technology Development and Information Company of India', the country's first venture capital financing initiative promoted by ICICI and UTI Infrastructure Technology and Services Limited. N. Sampath, who was heading the programme planning group at SAC moved in as chairman, Technology Transfer and Industry Interface.

A great soul (now in his heavenly abode), Sampath had been a caring elder brother since I met him.

Jai Singh was then preparing to take up an assignment with the International Maritime Satellite Organization (INMARSAT). He ensured a seamless transition of leadership of the Satellite Communication Programme. The baton was handed over to K. Narayanan from SAC, a 'Padma Shri' awardee by then. I must admit that it took me a decade to decipher the humane side of Narayanan, as outwardly he looked like an extremely reticent individual.

Rajan Sahib, an outstanding scientific secretary, ISRO, since 1978, was also the director of Earth Observation Systems and the secretary of the National Planning Committee of NNRMS. A few central agencies in the S&T system at Delhi had been persistently pressurizing him to come aboard since 1985. Finally, it happened in June 1988. He was appointed as the member-secretary of the newly constituted high-profile Technology Information, Forecasting and Assessment Council (TIFAC).

Besides elevating Srinivasa Setti, the then assistant scientific secretary, Prof. Rao brought in M.G. Chandrasekhar (MGC), the mission director of IRS-1A, to take up all three roles that Rajan Sahib had earlier handled. Without swaying amidst the ripples and the discontent in some quarters, MGC settled down at the headquarters.

Prof. Rao's intuitive and tactical counsel brought MGC and me to the same boat. We shared this bond for quite a few years.

With the hat of director, BEA, I became one of the five most senior directors at the headquarters, along with K. Narayanan, N. Sampath, MGC and D. Narayanamurthi. A new working relation and processes evolved at the headquarters.

In June 1989, Prof. Rao reiterated his faith in my abilities, appointing me as the director of RRSSC. This meant leading a team of 150 professionals, including 100 scientists and engineers. It was a bit unconventional, since as director I was one grade below four out of the five heads of RRSSC! That was where mutual professional respect, trust and interpersonal skills got tested. We, the five senior colleagues, worked as a cohesive team and met several professional challenges. Leading the five RRSSCs became my prime role. But I still continued as director of BEA.

One of the main tasks at hand at RRSSC was to push forward 'Vasundhara', a project on geospatial mapping for the entire southern peninsula, taken up at the behest of GSI.

By then, the Planning Commission had thrown a tough demand on the remote sensing community—to provide information on agricultural land use of every district in the country, within two years.

I sensed it to be the best opportunity for the digital technology to wade in. While naysayers opposed it and one of them even termed it blasphemy, Prof. Rao blessed our aggressive efforts and stood by us. Dr Karale and his team demonstrated accuracy, reliability and viability of stratified digital classification to generate this information.

The naysayers accepted it after a series of reviews were held at RRSSC, Nagpur. The agricultural land use information for 168 districts provided by the RRSSCs during 1989–91 became a milestone for digital image processing in the country.

My hard work got the recognition it deserved. I was considered an exceptionally meritorious case and was promoted to the next grade, six months ahead of the normal residency period (i.e. from 1 July 1990). This decision was taken after a gruelling promotion interview.

With this out-of-turn promotion, I had to walk the extra mile to meet the high level of expectations and also impose self-discipline to avert complacence. I managed to carve out a meaningful role for the RRSSCs in remote sensing applications and software development, complementing, supplementing and occasionally competing with the NRSA and SAC.

At times, even the large-hearted veterans such as Prof. Deekshatulu and Dr George Joseph would get irritated with my occasional aggressive behaviour. We accomplished several feats collectively, be it the proposal for setting up remote sensing data acquisition and processing systems for specific customers or enhancing the scope of the integrated geospatial study for watershed management. There was very healthy professional competition among units at ISRO.

I was tasked to acquire the new generation IBM workstations (forty of them) and software, working on a crash basis, as export restrictions had just been relaxed for ISRO.

The moment when I reported to Prof. Rao with the news that the systems had arrived at Bengaluru, on his last working day, 31 March 1994, was charged with emotion.

10

MADE FOR EACH OTHER

At about 8 p.m. on 31 December 2014 the calling bell rang. Our beloved niece Dhanya and her spouse Dhruv had arrived. Their sombre faces lit up as I opened the door. They presented their Cheriachan and Meemi (as Dhanya fondly calls us) a set of Carnatic instrumental music albums, a gift on the day of my retirement.

As always, Mini's eyes sparkled as Dhanya walked in. Dhanya has shared a strong bond with Mini since her childhood.

Just after our wedding function, many were amazed to see Dhanya, just one-year-old at the time, in my arms. In a way, I got Dhanya's endorsement on that day. Maybe, providence too had a grand plan that was waiting to be unveiled.

Mini's elder sister (Dhanya's mother) Radhika was married to C. Viswanath, a manager in Bank of Baroda, a good friend of my brother and a colleague of my first uncle. Viswan and I got along and eventually became close friends; he had become a favourite of my father's too. A decade after my wedding, on 22 August 1993, Viswan shocked us all and left us forever at the age of forty-two.

Dhanya had to be relocated to Bengaluru, where she continued her schooling; she went on to finish her engineering degree and later a post-graduation in management.

We supported Dhanya when she chose Dhruv, from Bhopal, as her life partner; he was her batchmate and colleague at a later stage. We cherished the new relationship with the respectable family of Vinod Sohanlal, a retired assistant general manager of State Bank of India, known for his uprightness.

They got married in the typical Kerala Nair style in Bengaluru on 26 December 2011. The next day, I had the rare experience of escorting the bride to the marriage blessing ceremony, held at St Mark's Cathedral as the Sohanlals were devout Christians.

The Girl Next Door

Dhanya, Dhruv and Nilanjan settled around the dining table. Mini got busy making crisp dosas; I carried these out to the youngsters' plates.

Dhruv remarked, 'They are made for each other.' Nilanjan agreed, 'True, a wonderful combination.' Dhruv interjected, 'Cheriachan, how often did you meet her before your marriage?'

I said, 'I first saw Mini in April 1983 during a circumambulation ceremony at the Koodalmanikyam temple in Irinjalakuda.' Nilanjan asked suspiciously, 'What took so long for the two neighbours to meet?'

I replied, 'It is true that our ancestral houses were just 500 metres apart. But, Mini spent most of her childhood in Thrissur as her father, V.P. Menon, was an engineer there in the Kerala State Electricity Board. They moved to their ancestral house in 1971, but by then I had left for Thumba.

'After her father succumbed to heart ailments in 1972, Mini's uncle, Raghavan Menon, took over the responsibility of their ancestral house. She continued to live there with her mother, Visalam Menon, and two siblings. Truly a noble man, he was a manager with State Bank of Travancore and a connoisseur of traditional arts. I too knew him since my childhood, as a good friend of my father's.'

I continued, 'I used to religiously participate in the annual festival at the Koodalmanikyam temple. In fact most of us from

around the precincts of the temple used to assemble there for
the ten-day-long festival. I gained popularity among the younger
generation after a couple of stage performances in the early
1980s at the temple. Mini's younger brother, Muralimohan,
and cousin, Appan, were prominent faces among them. We
used to flock around the percussion ensemble and Kathakali
performances.'

Mini added, 'In 1982, many of my classmates and friends
were enthusiastic to watch him dance at the temple festival.'

Dhruv was restless. 'Were you also among them?' he asked.
To which she firmly replied, 'No.'

I continued, 'In 1983, when I landed up for the festival at
the temple, one of our common friends, K.V. Chandran, told me
about Mini.

'Many ardent devotees from the neighbourhood would join
the first four rounds of the procession. The deity would be carried
atop a caparisoned elephant. As I spotted her in the crowd, I knew
that she was the woman I would marry. I gave a green signal to my
friend Chandran.'

All three smiled. These youngsters had got the answer they
were looking for. They soon left, and Mini and I retired for
the day.

Since we were again starting a new phase of life, memories of
those days came pouring back.

Building a Life Together

On 6 May 1983, I was in the middle of a conference at the ISRO
headquarters when the telephone rang. The operator said, 'Mr
Radhakrishnan has an urgent trunk call from his home.' My uncle,
Balakrishna Menon, was on the line, and he said, 'A proposal has
come for you. The girl is Raghavan's niece, Mini.' My response
was spontaneous, 'It is okay with me. You can go ahead.'

There was a long pause on the other end of the line. He was
well aware of my response to all the proposals that had come over

the past six years. They had seen me wearing a *rudraksha mala* and sporting a long beard for the past few years.

My uncle said, 'You should come down to see the girl formally at the earliest.' I agreed to reach Irinjalakuda on 9 May itself. I had to travel from Bengaluru to Thumba for a Project Management Board meeting and could easily stop at Irinjalakuda on the way.

Our families had been acquainted with each other for over three generations. Our mothers were classmates in school and friends too. Mini had been my mother's student at high school.

A first-class graduate in chemistry, Mini was employed in State Bank of Travancore; posted at Irinjalakuda. My father had professionally known her for a couple of years, as he used to do valuation of landed property pledged for loan from their bank.

As soon as Chandran hinted about the proposal, my father rushed to Mini's house to collect her horoscope; our common family astrologer certified it as an 'excellent' match.

It was a simple morning wedding at the Guruvayoor temple on 1 July 1983, followed by an exchange of garlands at Koodalmanikyam temple, and a brief function at the adjoining Unnayi Warrier Smaraka Kalanilayam hall. Several of our elderly relatives, teachers and friends participated and blessed us.

G. Balagopal (Bose), our English teacher Gopala Menon's son and childhood friend of both my brother and I, flew down from Kuwait. He and another close friend from Bengaluru, Padmanabhan (Pappan), took colour photographs of the affair, a luxury in those days.

Later, there was a reception with high tea for common friends in a hall at the National High School.

A month passed quickly, as we visited our close relatives, followed by a short trip to Mysore (now Mysuru) and Bengaluru. Soon, it was time to return to our respective workplaces in Bengaluru and Irinjalakuda.

The next three months seemed to have no end. The only options we were left with were writing letters, frequent trunk calls and a few occasional trips. After four months of effort and a little help from Mr Seshan, Mini finally got transferred to the Bengaluru City branch in Chamarajpet.

In December 1983, we settled down in rented accommodation next to the Srinagar bus stand. It was just four kilometres from her bank, so she could commute with ease.

The memories of the next thirty months are still fresh. Setting up the household, getting cooking gas and kerosene and struggling for water were the challenges we faced. Mini acknowledges that I taught her how to cook. She is a strict vegetarian and I decided to follow suit, at least at home.

We would never miss our evening strolls along the adjacent Ramanjaneya Road and the Kumaraswami temple Road that had an array of shops on either side. We used to religiously visit the nearby Balaji temple. We were able to attend most concerts held during annual music festivals as the Fort High School and the Gayana Samaja were located near the bank.

However, the cold weather of Bengaluru, which was very different from the hot and humid climate of our home town, had an adverse effect on Mini's health for the next three years.

In June 1986, we moved to Jeevan Bhima Nagar housing complex of ISRO, which was home to 125 closely knit ISRO families. Close family friends such as Pappan and Shailaja, Kutty and Valsala, Ravindranath and Saraswathi auntie along with their loving children Krishnan, Suja, Susi and Suni took good care of Mini, especially when I was away on official tours.

Dr P.N. Bhat was the custodian of our health, as he was for the rest of the ISRO families there. Dr (Mrs) Dhingra soon joined our ISRO family. Mini was impressed by our neighbour Santhi, and chose her to be her younger brother Muralimohan's life partner.

There were ample opportunities for sociocultural activities. I restarted my training in Kathakali and also did a few stage performances. In the vacant parcel of land just behind our quarters, a few residents renovated a Siva temple; I presided over the renovation committee.

Mini possessed several traits that complemented and supplemented mine. I must admit that, on many occasions, she proved to be a better judge of people than me. She has always been my best and most trusted critic.

She discontinued her music lessons after her father passed away, yet she had an impressive grasp of Carnatic music. Often, she would point out and correct my mistakes during my practice sessions at home. She taught me the composition, *Bhavayami Raghuramam.*

I used to be amazed by her skills in handling electronic gadgets which I never tried myself, even though I have an engineering background. In fact, she had better inputs to offer when we sat together in 1993 to plan our house on an undulating parcel of land in Banashankari.

Difficult Decisions

Even after several years of marriage, our family did not expand. Medical reports showed that we were truly made for each other. Both of us appreciated each other's will power to face the negative revelations. Recourse to modern medical interventions or legal adoptions was not in our scheme of thinking. We believed that we had been blessed lavishly in several ways; hence we resolved not to waste the rest of our valuable lives brooding over this one limitation. The associated social stigma was there for quite a few years, nevertheless.

A devotional poet of the sixteenth century, Poonthanam, wrote in *Jnanappana* (the song of knowledge), '*Unnikrishnan manassil kalikumbol, unnikal mattu venamo makkalai.*' This translates into: 'Do we need little ones of our own, when little Krishna is dancing in our hearts?' Maybe this helped me forge a strong connection with the younger generation.

We moved on and found numerous young people and little ones who found special places in our hearts. A few among them left their indelible imprints on our lives over the next two decades. The love of these caring ones raised our energy levels manyfold, and they became virtual members of our family.

Mini had a good reputation in her bank ever since she had started working at the Irinjalakuda branch. In the normal course, she would have easily gone up to the level of assistant general manager by the end of her career. But her priority was to be my emotional anchor, and she happily chose it over her own career.

She decided to forgo her chances of a promotion to the officer level as that would have required her to move around. Second, she had already taken an interstate transfer in 1983. This meant that she could not get a promotion at the new office for at least five years.

In August 1997, she was up for a promotion and a posting, but as fate would have it I got transferred to NRSA, Hyderabad. She didn't even think twice before letting go of that opportunity. Her transfer to the Hyderabad branch at Koti materialized by the summer of 1998. As per the guidelines, she became the most junior member of the staff there. But she remained patient even in that suffocating ambience just to provide me succour in an excruciating period of my career.

When I moved to Thiruvananthapuram in 2007, she accepted the loneliness at home as she took a long leave of two years to join me. Finally, when I came to Bengaluru as chairman, she decided to take voluntary retirement with effect from March 2010.

Thereafter, she immersed herself in household chores and taking good care of me, leaving me with no worries about managing domestic work.

Mini always stood by our close relatives and friends. She never hesitated to be their confidante and counsellor even when it affected her health. Wealth never fascinated us; rather we found contentment in helping the needy. We believed that it was our duty and an opportunity to extend a helping hand to others, without any expectations.

Mini was and still is my pillar of strength. Our similar upbringing, sociocultural and spiritual outlook, and a shared value system helped us swim together in both docile and turbulent waters with sufficient aplomb.

11

THE REFUTATION AND REAWAKENING

Prof. Rao had an amazing run in his last two years as the chief of the Indian space programme. His numerous initiatives and some gritty decisions had started bearing fruit and this continued even after he had left the centre stage. The big success came when the home-grown ASLV-D3 orbited the SROSS-C satellite, weighing 106 kilograms. The impact of the success was huge as this had been the test bed for several advanced technologies that we were developing, and these technologies were to be incorporated into our future programmes as well. Moreover, team ISRO had undergone a couple of agonizing successive failures of the same rocket in 1987 and 1988. The success of ASLV-D3 thus meant a lot to the organization. Prof. Rao's tough persona softened as tears of pride rolled down his cheeks. It was as if life had given him his best gift right before his sixtieth birthday.

There was well-deserved jubilation all around. However, ASLV-D3 delivered a tad short on the targeted perigee. This was unacceptable to Dr Gupta, the then director of VSSC, and he got immersed with his team to identify the problem. The two minor issues in the last leg of the flight were soon pinpointed.

In July 1992, ISRO had another major success in the form
of the INSAT-2A satellite. INSAT-2A, like ASLV, was home-
grown and it soon joined the INSAT-1 satellites (built abroad) to
provide operational services in communication, broadcasting and
meteorology. A year later, INSAT-2B joined this constellation.
Buoyed by this string of successes, we at ISRO started preparing
for the first developmental flight of our new launcher, PSLV,
planned on 20 September 1993. Our aspirations did not meet a
happy end here. Three flaws, all happening in conjunction, led
the mission to a heartbreaking failure. In retrospect, the failure
of the first flight of PSLV was probably a blessing in disguise. It
prompted a revamp of the autopilot system and flight software
of PSLV. Today, everyone can see the result. The PSLV has
established itself as one of the most reliable launchers in its class
with only one failure in its history till date.

In January 1994, Dr Gupta, the director of VSSC and a
member of the Space Commission, superannuated. Prof. Rao,
too, was walking towards the twilight of his tenure. On a personal
note, two of my fatherly figures in the professional realm were
retreating to the backstage.

Change of Guard at Antariksh Bhavan

Prof. Rao took a bow and signed off on 31 March 1994, after
a decade of an eventful tenure. His beloved colleague Dr K.
Kasturirangan (Dr Rangan) took over as chairman of the Space
Commission, secretary of the Department of Space as well as the
chairman of ISRO. An impressive gathering was organized at the
upbeat Antariksh Bhavan, befitting the occasion. Prof. Satish
Dhawan graced the event, as a proud grand-chairman. Dr Rangan
had been the director of the ISRO Satellite Centre since 1991 and
a member of the Space Commission since 1992. It was, again,
another seamless transition at the top office of ISRO.

Of course a couple of senior functionaries were disappointed
and exited in search of greener pastures.

Prof. Rao retained the roles of chairman of the Antrix Board and the council of the Physical Research Laboratory (PRL). He adorned the first Vikram Sarabhai Distinguished professorship for the next five years and also continued in the Space Commission till 2001.

Dr Rangan's innings at the top had a sterling start as the ASLV-D4 successfully orbited the SROSS satellite on 4 May 1994, and this time it was a perfect launch. What followed was a glorious nine years of ISRO under his leadership.

I had had a fairly close working relation with Dr Rangan over the past decade and had known him ever since I stepped into the headquarters in 1981. He had had a terrific career and was the project director of India's first two experimental Earth Observation satellites—BHASKARA-I and II—as well as the IRS-1A satellite. He was only forty then. His conceptual clarity and comprehension of any scientific and techno-managerial issue were remarkable.

A few months later, I accompanied the new chairman to Sriharikota for the launch of ASLV-D4 and PSLV-D2. PSLV-D2 was to launch the IRS-P2 satellite weighing 800 kilograms. PSLV-D2 was a very important flight for us as the first flight had failed, and we believed that we had incorporated all the changes from the previous one. The month of October is usually notorious in terms of weather near Sriharikota's coast. As expected, a possible wind was forecast during the launch window that could create problems in the early leg of the flight. Our meteorology team put their heads together with the launcher team and finally the launch window was cleared after a week.

On 15 October 1994, the PSLV-D2 had a perfect flight, and that wonderful journey continues even today as the PSLV keeps clocking one resounding success after another. For me, it was a learning process as I observed a magnificent team at Sriharikota, giving their best under the guidance of several towering leaders and Mr Madhavan Nair, the mission director. The first project director of PSLV, Dr S. Srinivasan, was by then the director of

SHAR and chairman of the Launch Authorization Board. In view of the importance of this mission, Dr Rangan took all concerned in the country into confidence to prepare for any possible outcome.

With the success of both these launches, Dr Rangan took up intense engagement with the ongoing 'Integrated Mission for Sustainable Development' (IMSD). Dr D.P. Rao of NRSA was the mission director of IMSD and I was officiating as the member-secretary of the Mission Management Council. The council was being chaired by Dr Rangan.

Those days, RRSSCs had introduced digital platforms for a multilayer spatial database of watersheds and decision rules that could be handled by the functionaries at the grass-roots level to prepare action plans for several developmental activities such as groundwater exploration and recharge, surface water harvesting, soil and moisture conservation, alternate land use practices for sustainable development and so on. Also, Y.V.N. Krishnamurthy and the teams at several RRSSCs successfully demonstrated the efficacy of the method in Chandrapur district of Maharashtra.

Dr Rangan was impressed with the result, so was Mr T.K.A. Nair, the then additional secretary of the rural development ministry, and also the secretary, B.N. Yugandhar.

Dr Rangan made a scholarly presentation to Prime Minister P.V. Narasimha Rao on 8 August 1994 on ISRO's preparedness to deliver a spatial database for watershed development at grass-roots level. The reach of the IMSD was thus expanded to specified blocks and watersheds from 174 districts. It was a significant step forward. We were tasked with delivering a spatial database for 85 million hectares—nearly 25 per cent of India's geographical area.

In late 1993, I came across Mathewkutty Sebastian, a postgraduate from IIT Kharagpur, who joined the headquarters under PC-NNRMS. He became a close associate as we spent a couple of hours together every day on the official engagements like the analysis of the voluminous data from the surveys or the preparation for a forthcoming conference. We became close friends too and Mathewkutty Sebastian became dear 'Mathus' for both Mini and me. Our mutual admiration and shared outlook on many subjects strengthened this bond between us. He became a virtual member of our family. In fact, my father confided in Mathus during the last phase of his life. I believe that meeting Mathus at that point of time was truly a silver lining for me.

The year 1994 was a hectic one. Apart from the huge responsibilities of leading RRSSCs through the ambitious IMSD programme, I was also engaged, as the organizing secretary, in the fifteenth Asian Conference on remote sensing that was held in Bengaluru from 17–23 November 1994. Prof. Rao and Dr Rangan continued to guide me in the roles they had assumed since 1993. The then deputy chairman of the Planning Commission, Mr Pranab Mukherjee, inaugurated the conference that was attended by 320 participants from several countries.

Professional Upheaval and Personal Loss

RRSSCs were largely responsible for the success of IMSD, and so were the other remote sensing application projects that I was leading from the headquarters. The position I had held in BEA had enabled me to plan long-term at the headquarters. With all these developments, I was under the impression that everything was going fine, and I would soon be promoted to grade G (joint secretary level) on completion of the minimum residency in the current grade. But to my surprise, it was delayed by six months.

Initially, I took the jolt. But I was taken aback and perturbed as I heard a new philosophy of differential promotional prospects for scientists at the headquarters. This was probably backed by a lack of appreciation for the nature of work in which scientists at the headquarters were engaged. This did not make much sense since we had been told for years that the headquarters must have the best.

However, this triggered a process of reflection in me. I felt it was time for me to dispassionately introspect on my skills and shortcomings. Enlightenment struck, I decided to 'focus' to 'deliver' rather than remaining a mere facilitator.

I spoke to some of my well-wishers for confirmation. I realized my introspections were leading me to the right path. I made up my mind. It was high time for a functional rotation even if it meant a change of domain, and enhancement of academic credentials.

By mid-1995, I approached Dr Rangan with these concerns. He readily agreed to support my doctoral research and suggested that I approach Prof. Dhawan for his guidance. He also advised me to get out of my comfort zone and probably take up a new role in a major centre of ISRO. I am forever grateful to Dr Rangan for this advice that helped me turn over a new leaf and initiated my professional rebirth.

I considered a couple of openings where I could utilize my past experience. The first was to move to ISAC, Bengaluru, to lead the Programme Planning and Evaluation Group, and the second was to relocate to NRSA, Hyderabad, as the deputy director.

The latter appeared more challenging and attractive for professional growth. I had sufficient exposure through leading the RRSSCs, which had five regional hubs of remote sensing data analysis and applications. NRSA, then an autonomous society with a strength of 1000 employees, has been the premier national agency responsible for satellite data reception, processing and dissemination. It was also the national nodal agency for aerial remote sensing, operational remote sensing application projects and capacity building. Moreover, NRSA appeared to be a promising platform for career progression.

I discussed my analysis with Dr Rangan and he advised me to stay for a while and wait for an opportune time for the movement.

In the meantime, he desired me to put my heart and soul into the formulation of the Indian Space Vision and Programme Directions for 2000–2010 and the Ninth Five-Year Plan (1997–2002). I introduced a questionnaire survey and elicited inputs from 500 members of the Indian space fraternity, especially the younger ISRO scientists, focusing on their perceptions and aspirations.

The result of the survey enriched our traditional planning exercise, involving programme planning groups at the centres, programme offices at the headquarters and the ISRO council. The new plan formulation exercise adopted by DOS was lauded and approved by the Space Commission in November 1996. That was indeed a moment of elation.

But amidst professional fulfilment came a great personal loss. On 29 August 1995, while my father, my two siblings and I were having lunch, he suddenly asked me, 'Radhakrishnan, do you realize that today is your birthday?' His choked voice was quite unusual. He would often tell us, 'Celebration of your birthday is to pay obeisance to your mother for giving you birth.' On that birthday I did exactly that, as I performed my mother's last rites; she had left us for her heavenly abode a day before.

I was closest to her, and I am not sure how I would have responded in my normal state of mind when she was diagnosed with advanced liver disease. Perhaps the almighty, in all his divine grace, had come to my rescue. A few days before I came to know of her ailment, I came across a medieval devotional and philosophical text, '*Hari Nama Keerthanam*' in Malayalam that gave an interpretation of human life and death. I was drawn into the literature the moment I saw it, and I finished it overnight. After delving deep into the philosophy, my state of mind was such that I could face the rudest shock that any son could get.

During the last twelve arduous days of her life in the hospital, there were occasions that made me feel worthy to be called her son. I was trying my best to comfort her, suppressing my emotions. I was happy that she would be free of her pain and would enjoy the company of her beloved brother, who had passed away recently, in the other world.

On a New Mission

'I am available for you twenty-four hours a day.' This was Prof. Dhawan's response when I approached him for guidance for my doctoral research. He looked quite convinced about it and asked me to write three pages each on at least three possible issues that

I would take up for my research work. During this exercise, Prof. Dhawan had cautioned me, 'Do not design for disappointment.' Finally, we chose a study on Indian Earth Observation Systems as the focus area for my research. Just before moving to Kharagpur, Prof. Dhawan advised me, 'Don't do anything that I won't do.' These words of wisdom, in effect gospels, still drive me. I was and I continue to be proud to hold the distinction of being his last student.

One cardinal decision that I took about my PhD was that I must earn it the hard way from a reputed institution and not resort to any shortcuts. The thirty-minute interview at the Industrial Engineering and Management (IEM) department of IIT Kharagpur went well. I was selected and registered as a doctoral scholar in July 1996. The head of the department assigned me to Prof. Damodar Acharya—my co-supervising guide.

The Doctoral Scrutiny Committee of five professors suggested I complete the coursework from January–April 1997. I decided to shift to Kharagpur from January 1997. Mini had agreed to cope with the change and Mathus was nearby for help. I got engrossed in the coursework and tests with students half my age; I was forty-seven years old then.

A significant development took place on 24 January 1997. Mathus was travelling that night to Vienna for a preparatory meeting of UNISPACE III and I called to wish him bon voyage. The 'goodnight call' from Mini was also over, and I got immersed in my coursework. Around 9.30 p.m., a senior engineer from Bengaluru called me up to tell me about the impending transfer of MGC from the headquarters to ISAC as its deputy director. He further indicated that two younger colleagues were likely to step into his shoes as scientific secretary of the headquarters and director of Earth Observation Systems.

I was curious about my move to the NRSA and asked him directly, 'Anything else?' He had nothing more to say. I was apprehensive, shocked and disheartened. I had proposed my move to the NRSA a long time ago. Even if this transfer didn't come through, I was still next in line for both these positions that MGC had held. I felt abandoned and clearly sidelined. The solitude of the guest house at Kharagpur allowed me to cry like a baby.

The next week, I went to Bengaluru and learnt about all the new developments. I had all the reasons to be concerned about my future, but an organization has to take decisions that are in its best interests. I had an opportunity to talk directly to the chairman and I expressed my concerns to him. I learnt that the decision had been frozen.

I came back to Kharagpur. To get rid of my mental turmoil, I took long walks around the isolated cells and gallows of the Hijli detention camp, which were part of the IIT campus. But those icons of India's independence struggle instilled a new spirit in me and elevated my mood.

Within a few days, as a godsend, I received a letter from my father, well past eighty-six then. In his stylish and steady handwriting, he wrote me a letter that pepped me up:

I am glad to learn that you are suitably accommodated and have entered into your studies in the right earnest. I hope you will pursue the Course with devotion, dedication and concentration to its logical consummation and achievement.

I am also glad that Mini is managing things at Bangalore well and that she has taken up her profession in the right earnest.

Please try to communicate with us as often as possible and convenient.

Please look after your health. I wish you all the success.

Yours affectionately, Father (22 January 1997)

That was his last letter to me. A couple of days later, his health started deteriorating and within six months he passed away.

But that letter did the magic. With that small, blue piece of paper, my father had reset my self-confidence to an extraordinary

Srinjan la kuda,
22 . 1 . '97.

My Dear Radhakrishnan,

I am greatly pleased to
have your letter which reached me
on day before yesterday. I am glad
to learn that you are suitably accommodated
and have entered into your studies in
right earnest. I hope you will pursue
the course with devotion, dedication,
and concentration to its logical consummate
and achievement.

I am also glad to know that
Muni is managing things at Bangalore
well and that she has taken up her
profess in the right earnest.

Please try to communicate with
us as often as possible and
convinent.

Please look after your
health. Wish you all success

Yours affly,
Father
R.M. Iyer

A copy of the letter that the author's father sent him in 1997.

level. I was re-energized, motivated and wanted to make the most of each moment.

I studied hard and walked a few extra miles. I became an active participant in the fortnightly round-table debates organized at the School of Management. Academically, I stayed in the top bracket. The periodical section of the IIT library and opportunities to mingle with several research students from diverse disciplines ignited my creativity.

My research work too progressed simultaneously, within my eighteen-hour daily schedules. I could feel that my guide, Prof. Acharya, and the Doctoral Scrutiny Committee were slowly becoming convinced about the possibility of my thesis submission within the next two years. That happened. I submitted the thesis in June 1999.

A significant take-home was a renewed urge to learn and apply the learning. My ability to learn fast and apply it faster got enhanced, and this helped me throughout my career as I switched between several domains.

By the last week of April 1997, I was back to work at the headquarters but I was in limbo for the next three months; my new role was yet to be carved out. The blessing in disguise was that I could frequently take leave to visit my ancestral house to be with my ailing father. My father was a bit surprised with my long absence from the office, and deeply saddened as he thought that I was still waiting for a posting after the study leave. I had to prepare myself to face adversities both at home and at the office.

On 19 July 1997, Dr Rangan announced that I would be moved to NRSA as the mission director of IMSD at the level of deputy director of NRSA.

I was able to convey this news to my ailing father, who breathed his last the next week. It was his last delight and he managed to move his right hand mildly to bless me. He left for his heavenly abode on 24 July 1997. After a fortnight of last rites and rituals, Mini and I returned to Bengaluru.

I wanted to take charge at NRSA on 22 August 1997. Before the departure to Hyderabad, I had an excellent session with Dr Rangan, during lunch, at his official residence 'Vyoma'. He gave me some advice on how to adjust to the new system.

Incidentally, on 20 August 1997 there was a colourful function at Antariksh Bhavan to mark fifty years of India's Independence, followed by lunch for all employees and their families. I sang a patriotic duet as the trinities, Prof. Dhawan, Prof. Rao and Dr Rangan, were seated on the dais. This was a fitting finale for my sixteen years tenure at the headquarters with these luminaries. The same afternoon, I relinquished charge of RRSSC.

The next morning, I handed over charge of director, BEA, to H.N. Madhusudhana. I met and greeted every member of the Antariksh Bhavan before boarding the flight to Hyderabad. I was moved by the grim faces and the fact that several of them had tears in their eyes as I shook hands with them.

Mathus reassured me that he would take care of my family as Mini had to stay back till her transfer materialized.

12

THROUGH THE GRASS ROOTS
AND GROOVES

I landed at the portals of NRSA on 22 August 1997. I felt, according to the NRSA management's interpretation of the developments of the recent past, a stage had been set for my rehabilitation. I had been appointed as the mission director of IMSD (a responsibility that had been shouldered by the director, Dr D.P. Rao) and also as deputy director, commercialization, at NRSA. That was a newly created position to accommodate me.

The position of deputy director of remote sensing applications, which was also being shouldered by the director, Dr D.P. Rao, still remained vacant and he continued to look after it directly; the office complex and support facilities were 'reserved'.

My innings at Hyderabad seemed to be on a sticky wicket. Maybe it was just my apprehension. I had to accept the fact that transitions came with the initial trauma that I had to cope with to adjust in the new surroundings. I faced these initial hiccups with a smile as I was prepared. Many expressed happiness as I was seen to be on the right side of the director of NRSA during the annual ISRO council meeting that took place in October 1997.

My strategic plan of commercialization of Spatial Data and Services, with the acronym 'SPARC', received accolades in December 1997, but later went into a tailspin. My position had become precarious in the organization by then. Clearance for the provisioned plan of implementation did not come through, though a few serious attempts were made to revive it. I decided to give it a go as a serious academic pursuit with the conviction that patience pays off and that no knowledge goes to waste in life.

Nevertheless, I had one more role and a major mission on hand. As the national mission director of IMSD, I had to lead a national team of thirty-five project directors and 200 scientists. A team of ten professionals of NRSA was to assist me as the IMSD core team.

The first task was to generate a spatial database of natural resources and an action plan for sustainable development of land and water resources in the remaining watersheds included in the mission. Second, quality and credibility of the database prepared by non-homogeneous teams had to be ensured. This was our plan to complete the primary goal of the mission.

We needed to worry about the proper use of the IMSD inputs by the state governments. While IMSD was providing only the spatial database, it was left to the state governments to adopt it in their plans; funding had to be found for implementing these plans.

The results were expected to show on the ground after a few years. However, the premature and exaggerated claims regarding the dramatic impact of IMSD, propagated from some quarters in the early 1990s, had given DOS a bad name. In several instances, I had to swallow bitter pills, but that was part of the game. This had to be overcome by discussing the real scope, its limitations and assumptions at the outset.

I travelled extensively across the country, to sixteen states, held discussions with different strata of officials and conducted in situ reviews at many watersheds. Apart from the scientific community, I could rope in the administrators serving on deputation at DOS. Each state had its own challenges for using IMSD inputs and we arrived at state-specific models for institutional arrangement.

I shared my experience through a presentation at the ISRO council in October 1999, befitting the conclusion of the task undertaken. IMSD, later, was acknowledged as the largest remote sensing application programme globally, completed through a participatory process.

My stint in Hyderabad still looked unfavourable and bumpy at times. The destination seemed to drift like a mirage. I was advised to carry the cross for some more time. I did so, but the journey seemed too long.

Meanwhile, I could not accept an offer to move to the North-Eastern Space Applications Centre as its director which was to be set up in Shillong, specifically for the north-eastern region.

Teachings of the *Bhagavad Gita* and Ramayana come to the rescue in such trying times. I also read Phillip C. Mcgraw's *Life Strategies: Doing What Works, Doing What Matters* many times and internalized its lessons.

I sensed a new Radhakrishnan or K.R. version 2.0 was evolving from a sensitive person to a sensible one with enhanced inner strength.

Mr M.R. Kurup, a noble person, revered by me and a few thousand others from Thumba, had been my well-wisher, more so after his retirement from the post of director, SHAR, in 1989. One day, in April 1999, I got a telephone call from Australia; Kurup Sir was on the line. He inquired, 'Is everything all right with you?' I said, 'Fine, sir.'

Kurup Sir did not seem to be convinced. He added, 'I am planning to return from Australia soon as my health is not permitting my stay here. Meet me in Thiruvananthapuram once I am back.' I met him at his residence there in June 1999. Kurup Sir looked weak. He was brief and asked me three pointed questions. At the end of the short discussion, he said, 'You also think about moving out.' Those were his last words to me as he passed away within a month.

In the autumn of 1999, two key persons from Antariksh Bhavan conveyed to me that my name was being considered for the post of director of the Indian National Centre for Ocean Information Services (INCOIS). This was a new autonomous institution being set up in the Department of Ocean Development (DOD).

Dr Muthunayagam, the founder of ISRO's propulsion systems and one of the most senior officials in ISRO's top management team then, had plunged into the oceanic realm in April 1995, as the secretary of DOD. He had already created a reputation by bringing in a paradigm change in DOD by introducing ISRO's way of working.

I felt if I was considered for the position, it would not be a bad idea to plunge into the ocean and swim with him.

Amidst these trying times, my PhD thesis reached an excellent shape. The thesis was submitted in July 1999; two evaluations, both positive, arrived by December 1999. I completed the thesis defence successfully on 22 February 2000 and became Dr K. Radhakrishnan.

I started preparing myself for yet another shift. The search-cum-selection committee held a personal discussion with a few candidates, including me, on 25 February 2000, at Delhi. I could present a comprehensive and focused plan for INCOIS. By the first week of July 2000, official intimation regarding my selection for the post of director, INCOIS, came through. I was assigned the grade of Scientist 'G' by the government that I had held at ISRO for the past five years. The level of the post was secondary to me as I was eager for a challenging task to prove my worth.

In the meantime, my case was not taken up for elevation to grade 'H' in ISRO at the end of the prescribed residency, i.e. five years as was the norm in those days (it's three years currently). That was the last nail in the coffin. I had, by then, realized the subtle differences between reaction and response.

On 6 July 2000, after twenty-nine years of dedicated service, I bid adieu to ISRO with a heavy heart, moist eyes and a choked throat. Procedurally, I had to take a 'voluntary retirement on immediate absorption basis' to move to DOD. I became a pensioner from ISRO, with the perennial financial disadvantage of a pro-rata pension for the rest of my life at the level of Scientist 'G'. I was reconciled to it.

But a grand design was to unravel in the years to come.

13

SAILING THROUGH THE
HIGH SEAS

'Life begins at the end of your comfort zone . . .'

I took over as director of INCOIS on 7 July 2000. Indeed, it was an excellent platform to perform, irrespective of the huge challenge of yet another switch of domain. Embracing such a unique opportunity marked the dawn of my professional resurgence.

Before making a confident plunge into uncharted waters, I carefully read Ian S. Robinson's *Satellite Oceanography: An Introduction for Oceanographers and Remote-Sensing Scientists*. Several aspects of the ocean are best understood from space platforms and that was the point of convergence between my past and present domains. Of course, there was a lot more in store to learn and practise about ocean observation, science and modelling.

INCOIS was the brainchild of Dr Muthunayagam, who moved out of DOS/ISRO and became the then secretary of DOD. The maritime satellite information system was being spearheaded by the erudite Prof. B.L. Deekshatulu, the then director of NRSA. A. Narendranath, a passionate oceanographer, incubated it within the premises of NRSA. I inherited it in this pristine phase, and that gave me an opportunity to guide the direction of its growth.

Shaping a Scientific Institution with a Difference

I started in my new role by forming a nucleus with two research scholars, Basant Kumar Samale and Thrivikram Prasad. K.K.V. Chary, who accompanied me from NRSA, and R. Devender, a talented casual hand, handled the administrative tasks. Three scientists allotted to me from DOD also helped for a short period. An ardent practitioner of the ISRO culture, I believed (and I still do) that a national institution should exist only if it can make a difference to the country.

By definition, INCOIS was meant to fill the existing void in the national information repository on maritime data, and hence it had a huge task at hand as our country has a coastline of more than 7500 kilometres. It also had global potential as information from the Indian Ocean was vital to almost all economies from south Asia. But, INCOIS had to strive hard to carve a place for itself among the established global players in the domain of ocean information.

'Unity is strength' is an old adage. It is evident that institutional networking and uniting strengths to achieve a focused goal will have more impact than several institutions struggling for one-upmanship. Obviously, one needs to inculcate institutional trust and team excellence to sustain the ambience of synergy. This holds good for all global initiatives involving common interests like space and the ocean.

India was evolving too. The emerging corporate culture of the late 1990s, especially the IT world, had several lessons in terms of compactness, efficiency and customer-centric delivery systems. And ISRO was one of the foremost government agencies to look for imitable organizational culture, systems and practices.

My first step was to define the vision and mission statements and bring out a strategic plan for INCOIS. We articulated the first vision statement, 'To emerge as an information technology enterprise for the oceanic realm.' But an admirable distinguished scientist, Prof. Vinod K. Gaur, had an amazing insight; he advised me to enhance its scope and add 'knowledge' as the other keyword in the vision statement.

The statement of our mission was 'to provide the best possible ocean information and advisory services to the society, industry, government and scientific community through sustained ocean observations and constant improvement through systematic and focussed research'. The cornerstone of the mission statement of INCOIS was synergy and knowledge networking with centres of excellence in ocean, atmospheric sciences, space applications and information technology as well as translating this scientific knowledge into useful products and services. In September 2000, the governing council, chaired by Dr Muthunayagam, endorsed the vision and mission statements and the strategic plan proposed by me, explicitly noting that it exceeded the original mandate of INCOIS to be a data and information provider.

During the first governing council meeting, a senior scientist wanted to test me and quizzed, 'What are the priorities of INCOIS in regard to research in marine biology?' My initial preparation and intense interactions with a few eminent marine biologists in the recent past helped me articulate the answer. That prompted an elated Gaur to state, 'You could now give mid-course correction to the marine living research programme of DOD!'

The strategic plan was comprehensive from scientific, technical and managerial perspectives. A proposal for a significant revision of the architecture of the technical and computational facilities was accepted. I was given the responsibility of inducting the start-up team and keeping the quality standards intact. The story of the five Pandavas and the hundred Kauravas, as told in the Mahabharata, played an important role in the hiring process. I focused on the talent of the recruits, rather than the number employed.

Soon, Dr Muthunayagam decided that INCOIS would, henceforth, handle projects on in situ ocean observations, satellite oceanography and ocean modelling, much to the chagrin of those who had administered from Delhi in the past.

One of the revolutionary advancements in those days for ocean observation was the concept of profiling Argo floats that enabled periodic measurement of temperature and salinity levels through the upper 2000 metres of the ocean. An international collaborative programme was evolving with the sponsorship of three UN bodies (WMO, IOC and UNEP) to realize a global array of 3000 Argo floats by 2006 for better understanding of ocean circulation and enhancement of climate predictability. India's commitment to deploy 150 Argo profiling floats in the north Indian Ocean was on the anvil, and that facilitated the foray of INCOIS into the international scene with a bang.

By the end of January 2001, Dr Muthunayagam completed his tenure as secretary. A renowned geophysicist, Dr Harsh K. Gupta (the then director of the National Geophysical Research Institute, or NGRI, at Hyderabad), took over as secretary of the Department of Ocean Development. After a few months of keen observation and understanding, Dr Harsh Gupta and I became an excellent team and our efforts were celebrated nationally and internationally.

Continuing at the NRSA campus as a 'paying guest' would have been detrimental to the institutional image of INCOIS. From the beginning I had advocated the shift of INCOIS to the Cyber Gateway that had come up at the HITEC City in Hyderabad. To me, the priority then was to make an impact as an institution rather than diverting our precious energy on development of a campus on land allotted in a difficult area. This proposal was shelved after protracted efforts.

In May 2001, Dr Harsh Gupta was convinced of the merit of my proposition of an independent campus for INCOIS. He told me, 'Radha, we are neither moving to HITEC City nor are we going to stay at NRSA. You function from a rented accommodation in the city till our own campus comes up. Also, let us develop the permanent campus on priority.'

He drafted Girish Pillai, director of administration in DOD, to team up with me and relocate to rented accommodation. We did so with due diligence, and within three months INCOIS started functioning from a decent building atop Nandagiri Hills, a prime location adjacent to the Jubilee Hills and the sprawling K.B.R. Park in Hyderabad.

By then, our start-up team of eighteen proficient professionals, with an average age of twenty-six years, was in place. I was fifty-two years old by then and donned the mantle of mentor and coach to the youngsters; half of them on their first job. Five of the seniors were entrusted to lead the key scientific portfolios.

M. Ravichandran, a brilliant physicist with research background from the Indian Institute of Tropical Meteorology and the National Institute of Ocean Technology, was entrusted with ocean modelling and ocean observations. B.V. Sathyanarayana, a postgraduate in computer science with rich experience in NRSA, took charge of computational facilities and web-based services. T. Srinivasa Kumar, a postgraduate in marine biology and oceanography headed advisory services and the satellite oceanography group. He had honed his skills in satellite oceanography and geographic information systems at ISRO and was a live wire at RRSSC, Kharagpur. Biogeochemistry was T.M. Balakrishnan Nair's forte, and he had put his heart and soul into the research of this new field at the National Institute of Oceanography (NIO). He agreed to lead the information services and ocean sciences group. E. Pattabhi Rama Rao, an atmospheric scientist with research experience from NIO, stepped in as the head of the data acquisition and information management group. I felt proud when even the youngest ones—Mohit Arora and Nagaraja Kumar—made a mark.

Girish Pillai and Praveen Tiwari, director (finance) of DOD, also became integral members of the team, *de facto*.

Through an institutional arrangement with DOS, we also drew on the expertise of Shailesh Nayak, Y.V.N. Krishnamurthy, P.G. Diwakar from ISRO, and A.S. Manjunath from NRSA for computational infrastructure and recruitment of the team.

What the Country Needed

In June 2001, Dr Harsh Gupta chaired his first governing council of INCOIS, held in NGRI. Several seminal decisions impacting the future of INCOIS were taken. More importantly, he fed me a piece of sweet from his plate at the end of the meeting. He was eloquent on the clarity and diligence of each proposal. He was amazed to see additional secretary and financial adviser C.S. Rao and joint secretary Anil Misra piloting the key proposals of INCOIS. Significantly, they jointly proposed that all delegable financial and administrative powers that the DOD enjoyed be delegated to its director.

I had met Prof. Dhawan before taking charge as director of INCOIS, to seek his blessings. He told me, 'You are old enough to realize that the science and technology system has a responsibility to serve the people of this country. What are you going to do to help the fishermen?'

In the 1990s, a process for identifying potential fishing zones using sea surface temperature derived from satellite data was developed by ISRO's Space Applications Centre at Ahmedabad. That had been scaled up by NRSA and maps were being faxed to fishery offices of the coastal states in the entire country as periodic advisories. INCOIS had imbibed this scientific heritage and continued it.

There was ongoing research at ISRO by Shailesh Nayak's team to include chlorophyll data from the Oceansat-1 satellite to improve these advisories and that was also ingested at INCOIS. The process of transfer of this know-how was not smooth as there was confusion at both ends. Antariksh Bhavan (DOS/ISRO headquarters) was not sure whether it should allow INCOIS to take it up, and Mahasagar Bhavan (DOD) was not sure whether it was worth-while for INCOIS to tag on to it.

However, the Tenth Five-Year Plan formulation process, led by Kalam Sir, proposed the Potential Fishing Zone Advisory Mission as DOD's flagship programme with DOS holding its hand.

In March 2001, INCOIS conducted a survey in three leading states for fisheries—Andhra Pradesh, Gujarat and Kerala—to get an insight into the ground realities. While there were no qualms about the scientific methodology, there had been several misgivings about the actual reach and utility of these potential fishing zone advisories. My initial task was to tweak the delivery chain to ensure that these valuable advisories became part of the decision-making process of the fishing community.

Srinivasa Kumar and his deputy, Nagaraja Kumar, established a robust operational system for generation of these advisories (during the fishing season, only for cloud-free areas) and translated them to local languages in an understandable format for the fishing vessel operators. We decided to deliver them directly to these end users along the entire coastline of the country. This had to be done regularly within fixed time slots.

INCOIS brought in a novel design of a digital display board to be hosted at the fishing harbours; this could be updated from Hyderabad regularly. The local fishing community was to host it. The initial trial at the Ratnagiri fishing port was a success. Soon, we started fishery information kiosks supported by Internet, mobile phones and print media to suit diverse profiles of fishing vessel operators on the west and east coasts.

An intense awareness campaign was started. I was thrilled to see the response at a dozen fishing ports where I interacted with the locals in Telugu, Tamil, Malayalam, Kannada and Hindi. We kept monitoring the usage through surveys and assessed economic benefits to the user communities. We found that the search time for fish had been reduced up to 70 per cent due to the usage of these advisories, and it yielded a saving of Rs 5,00,000–7,00,000 per year for each vessel operator.

We kept adding value to our information delivery. We added sea-state information to our bulletin. For safe operations at sea, the vessels needed to know about 'the state of the sea' or the wave height. We partnered with ISRO's Space Applications Centre and the Centre for Atmospheric Sciences (CAS) of IIT Delhi. Their research inputs were translated by Balakrishnan Nair as advisories on 'ocean state' forecasts and we disseminated them to the coastal states. The prime input for the forecasts was provided by the

National Centre for Medium Range Weather Forecasting, Delhi. Synergy made a difference here.

The website www.incois.gov.in was launched in the year 2001; it boasted a web-enabled geographical information system interface, data warehousing and data mining facility. Tata Consultancy Services (TCS) partnered with us in the development of the website. It continues to be one of the best ocean information websites globally. It was exciting to work personally for it, along with the young team of INCOIS led by Srinivasa Kumar and Mohit Arora as well as two veterans, Prof. Deekshatulu and Mr S. Vaidyanathan, who relived their younger days in the process. Diwakar, who was identified as the resource person from ISRO, passionately contributed and motivated the youngsters.

An International Responsibility

India is a founding member of the Intergovernmental Oceanographic Commission (IOC) of UNESCO, the apex body under the UN System on Ocean Sciences and Services, set up in the early 1960s. It has its headquarters in the imposing UNESCO building in Paris. In 2001, the Government of India decided to field a candidate for the position of vice chairman of IOC, and I was chosen as the nominee from India. I was a novice in the IOC system that had a legacy of four decades. The election was due in the twenty-first session of the IOC assembly scheduled on 13 July 2001.

My first appearance in the IOC circuit was from 28–30 June 2001, representing India in its subsidiary body i.e. the Intergovernmental Committee for the Global Ocean Observing System (I-GOOS). Dr Angus McEwan, a renowned oceanographer from Australia, was its chairman, and he was impressed by my interventions during the sessions. He was a key figure from the electoral group that had to back my candidature. Dr Harsh Gupta was heading the Indian delegation. It was a promising start, and

I made my presence felt in the IOC assembly during the next fortnight.

I was elected with acclaim as one of the vice chairmen. Along with the executive secretary, Dr Patricio Bernal, these six 'officers of the IOC' were to function as an effective team for the next two years (2001–03). I was re-elected as the vice chairman from 2003–05 and became the most senior vice chairman.

The International Argo Programme, endorsed by the World Meteorological Organization and IOC, was directed by an international Argo science team, led by Prof. Dean Roemmich of the Scripps Institution of Oceanography, University of California, San Diego. Along with Mr Stanley Wilson, the chief scientist of the National Oceanic and Atmospheric Administration of the US, Dean was looking forward to organizing the efforts of seventeen countries, including the UK, Australia, Canada, France, Iran, Japan, Malaysia, Pakistan and South Africa to deploy 450 Argo floats in the north Indian Ocean (north of 10 degrees south). India was contributing one-third of them.

At their request, INCOIS organized the Indian Ocean Argo implementation planning meeting at Hyderabad in July 2001. More than sixty delegates from seventeen countries (including Australia, Canada, France, Japan, the UK and the USA) and four regional organizations participated in the session. Dean's proposal to make me the regional coordinator was endorsed, and INCOIS became the regional Argo data centre for the region. Ravichandran led this effort with élan.

The Global Ocean Observing System (GOOS) is an internationally organized system for gathering, coordination, quality control and distribution of marine and oceanographic data, and derived products of common worldwide importance and utility. This has been a flagship programme of IOC. The most important means of implementation of GOOS is through the development of regional alliances which focus on issues of common national or regional interest.

Such a regional alliance was waiting to be forged in the Indian Ocean region as well. The Indian Ocean Principal's Meeting was held from 8–9 November 2001 in New Delhi to promote GOOS in the Indian Ocean region. It was decided to formally establish IOGOOS within a year and Dr Harsh Gupta was requested to lead this process. His magnanimity saw me being appointed as the chairman of the IOGOOS development committee.

During a dinner hosted on 9 November at the India Habitat Centre, I overheard Dr Angus McEwan from Australia telling Dr R.A. Mashelkar about my credentials for being chosen as the chairman. Dr Harsh Gupta asked me, 'Radha, how do you feel after scoring a hat-trick within just six months?' I said, 'Sir, I had to wade through a very bad patch in my career in the last few years. But god has been extremely kind to me.'

Soon, it was time for my thirtieth pilgrimage to Sabarimala shrine in Kerala, located in the Periyar Tiger Reserve of the Western Ghats. On all my previous trips, I had taken a shorter route of four kilometres to the shrine from the banks of the Pampa River. But, this time, as a mark of obeisance, I decided to take the most treacherous path of sixty-five kilometres from Erumely, through forests and on hill tracks.

IOGOOS was formally established on 5 November 2002 during the Indian Ocean Conference held at Mauritius. I was elected as the chairman of IOGOOS and continued in that post till 2006; INCOIS became the secretariat, ably run by Srinivasa Kumar.

Within a short time, IOGOOS emerged as one of the best GOOS regional alliances the world over. The most significant accomplishment of IOGOOS was facilitating the design of a unified observation system in the Indian Ocean for climate studies.

In March 2002, through an international review process, I was chosen as one of the experts to review the organizational structure of GOOS, to maintain its focus, efficiency and effectiveness in the changing times and circumstances. After conducting comprehensive surveys and inclusive reviews, the group of experts submitted its recommendations to the IOC assembly in 2003.

Subsequently, an open-ended working group of member-states, chaired by me, developed action plans that were adopted by the assembly in July 2005. This exercise was educative and satisfying.

The mandate on satellite oceanography and ocean modelling enabled intellectual interactions with the concerned groups from the Space Applications Centre of ISRO. These included the Centre for Atmospheric and Oceanic Sciences of IISc, Centre for Atmospheric Sciences of IIT Delhi, Centre for Mathematical Modelling and Computer Simulation of Council of Scientific and Industrial Research (CSIR), Indian Institute of Tropical Meteorology, National Centre for Medium Range Weather Forecasting, National Institute of Oceanography of CSIR, Naval Physical and Oceanographic Laboratory of DRDO, Andhra University and Cochin University of Science and Technology.

Nationally, that triggered a new phase of learning about the ocean fluid dynamics in space and time; mass and heat transfer between ocean and atmosphere; the unique features and complexities in understanding the Indian Ocean as compared to the Atlantic and Pacific and so on.

It was a moment of pride when these peers from the scientific community chose me as the prime mover for ocean modelling efforts in a mission mode, for achieving ocean predictability and enabling climate predictability with concerted efforts in modelling, data assimilation and validation through a network of institutions.

Distinguished scientists such as Dr M.S. Swaminathan, Prof. Rao and Dr R. Chidambaram professionally interacted with us, and soon became our patrons. I used to brief Dr Rangan personally and he had extended support from ISRO. He was visibly delighted as he stepped into INCOIS in 2003, and this symbolized the strong and positive interface between INCOIS and ISRO.

Moving Base

These were all the exciting developments when INCOIS was functioning from the rented office space at Jubilee Hills,

Hyderabad. During one of Dr Harsh Gupta's visits to Hyderabad, I handed him a document of my plan for the permanent campus, a task that he had entrusted to me in June 2001.

I felt that the building should reflect dynamism, singularity and convergence of different thought processes. I felt it should have the modernity of building concepts, modularity, provision for seamless expandability for the next thirty to fifty years, and my team members should feel at home there. He agreed and said, 'I have nothing to add. You plan for about 30,000 square feet of carpet area and present it to the governing council soon.'

My request to the Department of Space to help in the project management of the construction activity was accepted. Soon, we selected a suitable architect and a credible contractor to erect the campus.

When we presented the plan to the governing council, the financial adviser, D.P. Singh, commented, 'Doctor Sahib, who is going to be on the campus after 6 p.m.? You must add quarters for the scientists and the director.' Similar support came from his successors, Dr S.Y. Quraishi and Mr Arun K. Rath. Finally, the construction of the building, with a built-up area of 60,000 square feet and state-of-the-art technical support facilities, commenced in March 2003 with an eighteen-month schedule for its completion, i.e. August 2004.

It was completed on time and we started functioning from our own campus, named 'Ocean valley', in August itself. I knew every brick of that building. Dr Harsh Gupta too joined me at key points and brought in his wisdom of campus development. A paradigm shift in my thought process had already taken place.

The years I spent at ISRO still made me nostalgic, and I still shared an excellent rapport with my colleagues there. Every time I went to Bengaluru, I made it a point to drop in at Antariksh Bhavan for courtesy calls at the cabins of Prof. Rao and Dr Rangan.

On 27 August 2003, my beloved friend and former close associate at the headquarters, E.K. Kutty, telephoned me, 'There is wonderful news—Dr Rangan has been nominated for the Rajya

Sabha. The message came in just now. I just met him, he looked excited.' I called up Dr Rangan and congratulated him. I could sense the buoyancy in his tone. Within a few minutes, I made another call to a dear friend and colleague, Mr Madhavan Nair. I said, 'Sir, I pray that god blesses you.' He replied, 'We have to keep our fingers crossed. Anyway, I have been asked to reach Bengaluru today.'

After commendably steering the Indian space programme for about a decade, Dr Rangan laid down his office and was welcomed at the upper house of the Indian Parliament. Mr Madhavan Nair, the then director of the Vikram Sarabhai Space Centre (in secretary's grade) and a member of the Space Commission was given the additional charge of heading the Indian space programme from 1 September. It took three more weeks for him to be formally appointed as the chairman of the Space Commission and the secretary of the Department of Space and chairman, ISRO.

Mr Madhavan Nair and Dr P.S. Goel, the then director of the ISRO Satellite Centre, had both been in the grade of secretary and both were members of the Space Commission. But, seniority was respected in the appointment of Dr Rangan's successor at ISRO.

It was a great moment of joy to know about the function held at Antariksh Bhavan when Mr Madhavan Nair took charge and the first speech he made as the chairman. I took the next opportunity to come down to Bengaluru to personally congratulate my friend and the new chief of the Indian space programme.

Mr Madhavan Nair made it a point to visit INCOIS during his first visit to Hyderabad; it was a graceful reciprocation.

14

HURLED BY THE TSUNAMI

I was immersed in the *Anandabhairavi* raga while rendering the song *'Sabarimalayil thanka suryodayam'*, a particular favourite of mine, in a special music session organized at the Somajiguda Ayyappa temple in Hyderabad. Since 1997, I was staying in an apartment very close to the temple and was a regular there. The number delineated the *abhisheka* rituals for the deity at the Sabarimala shrine. I had completed forty-one days of austerities before making my thirty-sixth pilgrimage to the shrine.

I completed the rendition and opened my eyes. An endearing Laksmikiran, our neighbour and close family associate, whispered, 'Madam called up just now. You have to go home immediately. It looks like there is an official SOS.' It was supposed to be a calm Sunday after Christmas, but it turned out to be a turbulent one.

When I reached home, electronic media was already abuzz with the news of the rise in water levels and the substantial damage to the Andaman and Nicobar Islands.

The western coast of Indonesia and Phuket in Thailand had already been devastated. The risk was further snowballing as the furious waves approached the east coast of the Indian mainland via the Sri Lankan coast. It was expected to further advance towards the

African coast after engulfing the Maldives a couple of hours later. India was about to experience a 'tsunami' for the very first time.

Tsunami, a Japanese word meaning harbour waves, remained academic jargon for oceanographers of the Indian Ocean region till that day; although, it has been quite a frequent phenomenon in the circum-Pacific belt. This is the geographic region where more than 75 per cent of the earthquake energy is released. The countries around remained prepared for it with early warning systems and well-rehearsed protocols for rescue operations.

International coordination on the tsunami warning system in the Pacific Ocean came under the IOC. As the vice chairman of the commission, I had a bit of exposure to the functions of the Pacific Tsunami Warning Center at Hawaii, operated by NOAA of the USA, and Japan's Meteorological Agency in Tokyo.

It was soon unravelled that two tectonic plates under the Indian Ocean had collided and a deadly earthquake had struck the famous Sunda Trench near northern Sumatra in Indonesia that morning. The devastating tsunami was the result of this large-scale disturbance on the seabed. It turned out to be one of the strongest in the world and probably the deadliest of all times. The toll was close to 2,50,000 human lives in Asia (18,000 from India), besides indelible damage to the flora and fauna and huge financial loss. It even induced deformations in the topography. Humanitarian aid poured in from all over the world to supplement and complement national and regional efforts of rescue, relief, rehabilitation and reconstruction.

Establishing a Warning System

On 26 December 2004, in the middle of a meeting of the crisis management group in New Delhi, Dr Harsh Gupta called me up and said, 'Radha, if you have any relevant data at INCOIS, email it to me immediately.' I responded with the best I could.

It was a rare opportunity for the scientific community to join hands to serve humanity. We took a resolve to configure an early warning system for India and the countries around the Indian Ocean region. We planned it diligently during the next fortnight without making any public statements, unlike several experts who did.

Dr Harsh Gupta, a celebrated seismologist, convincingly established that there were only two tsunamigenic zones of the Indian Ocean—the prime one near Java-Sumatra in the east and the second one near the Makran coast in the west. These two locations fall within the Alpine-Himalayan belt where 20 per cent of the earthquake energy is released. These locations needed to be monitored continuously as earthquakes of large magnitude could trigger tsunami waves in the Indian Ocean.

All scientists of INCOIS were immediately pressed into the study of multiple facets of tsunamis and requirements for a reliable early warning system, also contemporaneous with technology. One of our elder sister institutions, the National Institute of Ocean Technology that had rich experience in ocean instrumentation, occasionally complemented us, offering healthy competition.

We figured out the tasks quickly. To detect intense earthquakes, we needed a chain of interconnected seismic stations and this was in the realm of the India Meteorology Department. To confirm whether a tsunami had been triggered by an earthquake, the water levels also had to be monitored using pressure recorders located on the seabed near the two identified tsunamigenic zones. Physical understanding and numerical modelling of the tsunami waves right from their source of generation was a challenge. All possible scenarios had to be generated in advance and stored; once an earthquake struck, the alerts and warnings could be issued instantaneously by picking up the relevant scenarios. Coastal radars and tide gauges near the coasts were essential to monitor the progress of the tsunami waves. Communication satellites with a footprint over the north Indian Ocean were essential for bringing all data to the warning centre. Knowledge on coastal topography and a spatial database of the coastal belts was required to forecast the possible areas of seawater inundation. Timely dissemination of warning to the coastal population was crucial.

We configured the system to cover cyclones, considering some of the commonalities. It is a well-known fact that 13 per cent of the world's cyclones originate in the sea around India; the east coast of India had experienced this frequently. We also realized that reliability of disaster warnings was important.

The early warning system had to be established in mission mode with all stakeholders on board. We aggregated national support from all scientific departments in the country. The secretary of the Department of Science and Technology, Dr V.S. Ramamurthy, extended support from India Meteorological Department (IMD) for seismic stations. The director general of CSIR, Dr R.A. Mashelkar, also came on board, ensuring support from NIO that opted to focus on tsunami-related research. Secretary, Department of Space, and my old colleague and friend, Mr G. Madhavan Nair offered communication satellite capacity and satellite-based terrain mapping.

By the second week of January 2005, the project proposal was presented to the Union science and technology minister Kapil Sibal. Soon, the home affairs ministry, responsible for disaster management, accepted this approach.

The Department of Ocean Development was mandated as the nodal department and INCOIS became responsible for early warnings on tsunamis. Thus a fatal day for Asia paved the way for a fateful day for INCOIS, the youngest oceanographic institution in the world then. As director of INCOIS, I got the opportunity to be the first project director of the Indian Tsunami Early Warning System. A versatile scientist and manager, Srinivasa Kumar, emerged among his peers as the anchor person to implement it.

While the Pacific Tsunami Warning Center (PTWC) at Hawaii or Japan's Meteorological Agency in Tokyo could provide certain information about tsunamis (earthquake information, a general statement about the potential for tsunami generation and estimated time of arrival at the coasts), an exclusive system located in the Indian Ocean was essential to confirm that a tsunami actually existed, to forecast its strength and for cancellation of warnings.

It was clear that such a system would suffice not only for India but the other countries in the entire Indian Ocean region. That opened up an excellent opportunity for India to establish regional leadership and gain goodwill as well as meeting her national needs. For the next five months, I literally lived out of my suitcase, in different countries around the world.

Involving the World

International deliberations on the tsunami warning system were taking place on a war footing basis since the first week of January 2005. A special meeting of the Association of Southeast Asian Nations (ASEAN) leaders was held at Jakarta, in which the decision to establish a regional early warning system was taken.

The second World Conference on Disaster Reduction (WCDR) organized by the United Nations was scheduled to take place at Kobe in Japan from 18–22 January 2005. This was being organized a decade after the first one, which was held in 1994. The Indian Ocean tsunami inevitably became an important topic. An exclusive thematic session and a special plenary session were introduced hurriedly. Dr V.S. Ramamurthy and Dr Harsh Gupta asked me to rush to Kobe, as a member of the Indian delegation led by the secretary of the home affairs ministry.

My presentation in the thematic session emphasized the Indian initiative for an early warning system that covered the two known tsunamigenic zones affecting the Indian Ocean region. We also made a commitment from the highest level in the government for implementation of the operational warning system by September 2007, at an estimated cost of USD 30 million. We also offered to contribute by disseminating warnings from our national warning centre to countries in the region. We proposed to utilize IOGOOS with its wide reach for this purpose.

Besides India, several oceanographic superpowers were keen to contribute towards the early warning system. These included Indonesia and Thailand from the affected Indian Ocean region, as well as other countries like Australia, China, Japan, Germany and the USA. Intense discussions at political and diplomatic levels took place till the wee hours at the Kobe conference. Just an hour before the session, Pankaj Saran, the then minister (political) of the Permanent Mission of India in Geneva, conveyed that I had been chosen to give a 'special talk' on the early warning system, as a representative of the Indian Ocean region.

This was a new topic of global concern and I had just been introduced to it! Indeed, it was one of the proudest moments of my life to be on the dais of the Kobe Conference for a seven-minute presentation before 4000 participants from around the world.

The outcome was a common statement from the participating countries at the conference. It has been produced verbatim below:

> Emphasizes that a tsunami early warning system for the Indian Ocean must be tailored to the specific circumstances of the Indian Ocean and the individual requirements of countries, under the coordination of the United Nations, and that those countries must be the ones to determine the shape and nature of the system.

The conference in Kobe was followed by a ministerial-level meeting on regional cooperation in tsunami warning arrangements, organized at Phuket by the government of Thailand. Ministers and special envoys from forty-three countries participated.

It was a privilege to be a special envoy of the external affairs ministry and to be a key player with the Indian diplomat T.P. Seetharam by my side. Our interventions were supported by Australia, Japan, Sri Lanka, the UK and the USA. This resulted in substantial changes in the draft declaration presented by Thailand to establish the Asian Disaster Preparedness Center as the

Regional Tsunami Warning Centre. The conclusion was to have a multinodal tsunami early warning system in the region. This arrangement was to be developed within the UN's international strategy, coordinated by the IOC of UNESCO.

Now, it was the turn of the IOC of UNESCO to forge arrangements and Dr Patricio Bernal, the executive secretary of IOC, became the anchor person. The first international coordination meeting for the development of an Indian Ocean Tsunami Warning and Mitigation System (IOTWS) was organized at UNESCO's headquarters in Paris in March 2005.

The Indian delegation, comprising officials from the Department of Ocean Development, Department of Science and Technology, and Department of Space, played decisive roles during the plenary sessions on the technical and organizational aspects. Also, I chaired part of the plenary session in my capacity as the vice chairman of IOC.

It was concluded to make IOTWS a coordinated network of national systems at two levels, viz. those with the capability to generate and issue warnings, and those who could receive and disseminate this information. An Intergovernmental Coordination Group for the Indian Ocean Tsunami Warning and Mitigation System (ICG/IOTWS) was recommended by the IOC to govern the system.

We also attended and played a key role at the second international coordination meeting held at Mauritius in April 2005. Several key countries from the Indian Ocean region expressed their support for India becoming the first chair of the ICG/IOTWS. The IOC secretariat at Perth, Australia, was identified to act as the secretariat to the ICG/IOTWS.

A spin-off from these meetings was that my candidature in the ensuing election, to be held in the IOC assembly from 20–30 June 2005 for chairman, IOC, was formally mooted by Dr Harsh Gupta. However, Mr David Pugh (UK) was extremely keen on a second term. The IOC rules did not permit two individuals to share the two-year term. As the Indian government did not want to stand against a nominee from the UK, I was advised to withdraw my name.

During the IOC assembly, David was re-elected for a second term as the chairman of IOC. This was followed by the election of the executive council of forty member-states where India was also contesting. I stayed back till the election, held on 27 June, as the Indian delegation felt that my physical presence in the assembly would make a difference to India's image. India secured eighty-six votes out of the ninety cast.

The international coordination was, hence, completed. We were well on course to have our tsunami warning centre and a leadership role among the global oceanographic domain. I returned to India in the last week of June 2005 with a great sense of achievements and goodwill.

I had to be in Bengaluru to represent DOD in the first meeting of the India-US Joint Working Group on Civil Space Cooperation being hosted at the ISRO headquarters. From India, the Working Group was co-chaired by Dr P.S. Goel, the then director of the ISRO Satellite Centre and also a member of the Space Commission.

Life was on the verge of a new turn!

15

HOME IS WHERE THE HEART IS

Dr Harsh Gupta was due to superannuate on 30 June 2005. The process of identifying a successor had started in January 2005. There were also discussions to transform DOD into the Ministry of Earth Sciences with an enhanced mandate.

I was the most senior among the three directors of DOD's autonomous institutions. Besides, INCOIS had the lion's share in terms of programmatic contributions at the national and international levels within DOD. I was, probably, the most likely successor to Dr Harsh Gupta internally.

For the past year, Dr Harsh Gupta had kept me involved in overarching matters of DOD, at times beyond the remit of INCOIS. Dr Satish Shetye, the then director of the National Institute of Oceanography (he was in the outstanding scientist grade), conveyed to me that he did not wish to be considered for the post of secretary and he felt that I should put my name for it.

However, from May 2005 onwards, Dr P.S. Goel from ISRO emerged as a likely candidate to head DOD. Dr Goel, a senior, had thirty-five years of experience in satellite technology. He was already a Padma Shri (in 2001) and carried very high credentials including fellowships from four premier academies in the country. To top it all, for the past few years, he had been in the apex grade, equivalent to the grade of Secretary to the Government of India, and he was a member of the Space Commission and the Scientific Advisory Committees to the prime minister and cabinet.

Obviously, Dr Goel was way ahead of me, if he really had been nominated. But, like all professionals, I was concerned about my own professional growth. I had completed nine years (including a few years at ISRO) in the grade of Scientist 'G' by then. I was fully aware that the concerted efforts of Dr Harsh Gupta and the then minister of Ocean Development, Kapil Sibal, to upgrade the posts of directors of DOD to the 'outstanding scientist' grade were not getting materialized. I was convinced that my career was about to get stuck unless new professional challenges came along.

Srinivasa Kumar (Srini), my trusted lieutenant in INCOIS, had become a close family friend. I consulted him for major personal decisions. Srini became somebody whom both Mini and I turned to in moments of crisis and confusion.

The next day, I took Srini for a long stroll in the INCOIS campus and said, 'I gather that Dr Goel is a likely candidate for heading DOD, and, if so, he would definitely make it. In that case, there might be a lot of lateral movements within ISRO. Director of SAC, Ahmedabad, Dr K.N. Shankara, might be brought into Dr Goel's place at ISRO Satellite Centre. Then Dr R.R. Navalgund, director of NRSA, might like to move back to SAC, his parent centre.' Srini responded with a smile, 'Director, NRSA, is a coveted post.' I got the cue. We did not dwell on it further.

It was official by the last week of June 2005. The Government of India issued an order appointing Dr Goel as the secretary of Ocean Development and he was to take up the baton from 1 July 2005. Dr Harsh Gupta, who was leading the Indian delegation in the IOC assembly (held from 22–30 June 2005) in Paris, was scheduled to come back to India the next morning. I was also with him in Paris as part of the Indian delegation and vice chairman of IOC.

Before he left for the Charles de Gaulle airport, we met for an early breakfast at Hotel Pullman Paris Eiffel Tower, where the Indian delegation was put up. Right then, I got a call from Antariksh Bhavan, saying, 'The chairman, Mr Madhavan Nair, would like to speak to you. Please call him up.'

After seeing off Dr Harsh Gupta, I called up Mr Madhavan Nair. The usual friendly warmth in his voice was palpable even over the phone. After brief pleasantries, he said in a low voice, 'Dr Goel is being appointed as the secretary of DOD. Now what are your plans?'

I didn't spare a second to reply, 'I'd love to come back to ISRO.' He promised to call me back shortly.

I returned to Bengaluru on 28 June and presented at the Antariksh Bhavan as the representative of DOD in the Indo-US Joint Working Group on Civil Space Cooperation, chaired by Dr Goel. By then, all our predictions about the movement of Dr Shankara and Dr Navalgund had come true.

Consequently, selection of director of NRSA, an autonomous body under DOS, was expected soon through an open advertisement. I spoke to Mr Madhavan Nair on my analysis of the situation and reiterated my wish to be a candidate for the post. I met both Dr Rangan and Prof. Rao. Both of them listened to my analysis of the situation carefully and blessed my intent to come back to ISRO. Several of my former colleagues and well-wishers at Antariksh Bhavan were visibly happy with the turn of events.

Dr Goel took over from Dr Harsh Gupta on 1 July 2005. I could not be present at Delhi since Mini had joined me in Bengaluru to celebrate our twenty-first wedding anniversary on the same day. I took permission from Dr Goel to stay back and he approved it. He knew us well as we were neighbours in Bengaluru for a short duration.

I formally called on Dr Goel in Delhi a couple of days later, and he made his first visit to INCOIS the next week itself. Many colleagues in Mahasagar Bhavan looked a bit surprised to see our camaraderie.

My close association with Dr Harsh Gupta continued, as he soon relocated to Hyderabad, close to the National Geophysical Research Institute.

DOS notified all the S&T departments about the recruitment process for director, NRSA, in July 2005. Dr Goel marked a copy to me. I meticulously prepared my application with the help of Srini and sent it to him with a request to forward it to DOS. After speaking to me to understand my view, he forwarded my application to DOS.

DOS did its due diligence with the utmost sincerity. It had sought nominations of potential candidates from former directors of NRSA. One of them, Prof. B.L. Deekshatulu, called me and asked, 'Do you want to wait for a few years and become the secretary of DOD or head NRSA right now? If you want to opt for the latter, meet me tomorrow at 10 a.m. with your biodata.' I thanked him, and promised to meet him the next day in his cabin in the Artificial Intelligence laboratory at the University of Hyderabad.

The next day, Dr D.P. Rao, the former director of NRSA, too asked for my biodata and recommended my nomination. I felt extremely happy at this gesture; life had come full circle for me.

My aspirations of going back to ISRO didn't affect the pace at which INCOIS was marching ahead. Dr Goel and I concentrated on organizing new technology elements for the tsunami warning system. In the process, I had to go to Perth, Australia, to represent India in the first meeting of the Intergovernmental Coordination Group for the Indian Ocean Tsunami Warning and Mitigation System. Again, within days, I had to be in Bali, Indonesia, to chair an IOGOOS meeting that had an important agenda of deliberating and adopting a report by an international panel of experts on unified observation systems in the Indian Ocean for climate studies.

Back home, DOS shortlisted five candidates for NRSA and invited all of them for personal discussions with a distinguished selection committee at Antariksh Bhavan on 6 August 2015. I was one among them.

Dr Goel graciously allowed me to return from Perth to Bengaluru for the personal discussion and then fly to Bali.

Preparedness had been my habit for any meeting; more so when it was an attempt to pluck a much-desired fruit. While I was oriented towards management by training, space technology and applications had been my professional forte for a long time. My PhD thesis on 'Some Strategies for the Management of the Indian Earth Observation System' was a straight fit for the role required at NRSA. I put together a twenty-slide Power Point presentation, delineating my vision for NRSA and the strategy to accomplish the same. I was satisfied with my preparation.

I entered Antariksh Bhavan with fond memories of my interview at VSSC thirty-four years earlier. It was indeed a warm welcome from all levels. Some of them even told me, 'We want you to come back.' I was touched by this expression.

The selection committee was chaired by Prof. M.G.K. Menon; Mr Aravamudan (former director of ISRO Satellite Centre) and Prof. Sulochana Gadgil of IISc were its members. A scholarly discussion took place. I was selected.

Both DOS and DOD worked in tandem to process the papers for seeking approval of the appointments committee of the cabinet. DOS started initiating me into NRSA informally. I recall with gratitude the briefing sessions with Dr Navalgund at Hyderabad and with Mr Madhavan Nair at Bengaluru on a few occasions.

My last serious engagement with DOD was the brainstorming session organized by Dr Goel in Delhi to formulate the Eleventh Five-Year Plan for the Department of Ocean Development. I contributed wholeheartedly to this session, conducted over two days. Distinguished scientists, including Dr S.Z. Qasim, former secretaries and directors were present. One of them, Dr S.A.H. Abidi, who was associated with oceanography in India since the 1960s hugged me and said, 'I will always be proud that I knew you and worked with you.'

I was touched, this was probably the biggest certificate that I got while working in the realm of ocean.

16

REUNION AT NRSA

'Buddy, welcome back! Congratulations. You have to be in Delhi tomorrow morning for your first meeting as the director of the National Remote Sensing Agency (NRSA).' The happiness in V. Sundararamaiah, the then scientific secretary of ISRO, was evident. It was a momentous day in my life too.

The joint secretary, R.G. Nadadur, coordinating personnel matters in DOS, had called me up the previous morning to break the news that approval from the appointments committee to the cabinet was on its way. Also, he had said that Mr Madhavan Nair wanted me to take charge at NRSA as soon as the order was issued.

In the afternoon of 7 November 2005, I received the order from DOS, appointing me as director of NRSA in the 'outstanding scientist' grade. Thankfully, Dr Goel and his colleagues at Mahasagar Bhavan completed all the paperwork and formalities in fast-track mode to relieve me from INCOIS the same evening. V. Sampath, my esteemed colleague from DOD, was ready at INCOIS to take over.

With a great sense of satisfaction and several emotions floating around, I left the INCOIS campus late that evening. The twelve-kilometre journey to NRSA passed in a flash. I was excited to reunite with my former colleagues and was filled with passion to begin a new fruitful innings at ISRO.

I arrived at NRSA at 7.45 p.m. to take charge. Several of my colleagues were waiting with warm and welcoming smiles.

Emotions poured in as I sat on the chair of the director; many well-wishers at NRSA had longed for this event in my first stint itself.

I recollected and reaffirmed to myself, *'Whatever happened in the past, it happened for good; whatever is happening, is happening for good; whatever will happen in the future, will happen for good . . .'*

Vijayachandra, my competent and trustworthy personal secretary for the next two years, greeted me with his familiar smile and handed over tickets for a flight to Delhi the next morning.

I landed at the Ganga International Guest House at Pusa in Delhi to be greeted by an elated Mr Madhavan Nair. I was so happy to be a part of his team. After the workshop, we discussed the imminent agenda of NRSA and a few general points on ISRO at large. One of the initial pieces of advice he gave me was, 'Look beyond NRSA as well.'

Within the next fortnight, I was required to participate in a two-day meet of the 'ISRO Programme and Policy Council' that took place at Mysuru. This council, a large group of nearly 100 senior functionaries of the top two tiers of the ISRO management, was the brainchild of Dr Rangan; Mr Madhavan Nair had nurtured the same. This meet helped me update myself and revive the interfaces.

I also became a member of the ISRO council, an apex body within ISRO, to handle techno-managerial matters. By 19 January 2006, I was inducted as director of the board of Antrix Corporation Limited, a public-sector undertaking under the Department of Space, in my capacity as director of NRSA.

My reinduction was complete.

Leading NRSA to the Next Level

As the chief executive of NRSA, I was tasked to lead a team of 500 scientists and engineers as well as 425 technical and administrative support personnel. These people were deployed in three different locations—the main campus at Balanagar, the Shadnagar Ground Station complex (about eighty kilometres away from the main campus), and the Indian Institute of Remote Sensing (IIRS) in Dehradun.

We arranged a befitting farewell for the outgoing director, Dr Navalgund; his two predecessors, Prof. B.L. Deekshatulu (1982–1996) and Dr D.P. Rao (1996–2001), were also felicitated. I took this opportunity to brief the gathering on the strategic plan that I had. The rapturous applause that followed reassured me of the enormous goodwill and support that I had.

The NRSA that I had left in 2000 had undergone phenomenal transformation. The imprint of Dr Navalgund's leadership was visible all over. A new and impressive entrance complex was renovated in the area that was once crowded with dilapidated buildings. The huge photo-processing machines had been replaced with modern computer workstations and video conferencing facilities. A switch to the digital domain for the satellite data delivery system was nearly complete. For me, the most significant change at NRSA was the attitude of the people, especially those in important positions. Many young and bright faces had risen to the second and third tiers in leadership, replacing several elders, who had earlier held these positions with discontent and frustration. I paid my compliments to Dr Navalgund for the evolution of NRSA.

It was my duty to make a clear assessment of the resources, responsibilities and risks of running NRSA as another centre of excellence in ISRO.

The Remote Sensing Data Policy (2001) had armed the NRSA with the sole authority in the country to receive data from Indian and foreign satellites, and process them into products and services that would meet the needs of a variety of users, subject to specific guidelines.

At that time, the IRS constellation comprised ISRO's second generation satellites—IRS-1C and IRS-1D, thematic satellites such as Oceansat-1 and Resourcesat-1, and cartographic satellites such as Cartosat-1.

IRS-1C (1995) and the identical IRS-1D (1997) had raised the global perception about India's capability on satellite-based earth observation. This was evident in the confidence shown by foreign customers from the US, Europe, West Asia and North Africa, who had entered into commercial arrangements to receive IRS data at their ground stations. It was NRSA's responsibility to make sure that these international customers received the data.

The launch of a versatile thematic satellite, Resourcesat-1, and a high-resolution satellite, Cartosat-1, by ISRO during 2003–05 put remote sensing applications on a new pedestal.

Resourcesat-1 was a versatile satellite with three complementary camera systems that scanned a large area once in five days, besides capturing finer details once in twenty-four days. This opened up new possibilities to study natural resources, especially agricultural crops and vegetation. The Cartosat-1 satellite could take images of a selected terrain at a spatial resolution of 2.5 metres (the best from ISRO till then) besides giving height information of the terrain. Oceansat-1 had instruments such as an Ocean Colour Monitor (OCM) and a Multi-frequency Scanning Microwave Radiometer (MSMR), which enabled oceanographic studies.

These satellites provided the required information for several national missions and projects on remote sensing applications that were undertaken by NRSA and ISRO, along with user agencies, to generate resource management information and value-added services in domains such as agriculture, water resources, geology, geomorphology, forestry and biodiversity. A few new initiatives for a natural resource census, natural resource database repository, infrastructure planning and terrain mapping followed.

NRSA was also operating two aircraft for close-up imaging of areas when required. Constraints imposed by an age-old process of mandatory clearances, at every step, were hampering the aerial services and related data dissemination, even for disaster management support. NRSA had a larger national role to perform. We needed to gear up.

At the first tier of managerial support, I had three deputy directors, a dean (at IIRS) and a general manager spearheading the scientific and technical entities.

S. Raghunathan, a veteran at NRSA since its inception, was the deputy director, in charge of the data acquisition domain. K.M.M. Rao, a soft-spoken engineer with novel ideas, was another deputy director heading the data processing area. A relatively young scientist, P.S. Roy, with accomplishments in remote sensing applications for forestry and biodiversity studies, moved in from the Indian Institute of Remote Sensing (IIRS) to become the deputy director of remote sensing applications and Geographic Information Systems (GIS). And, of course, we had the young and brilliant scientist, V.K. Dadhwal, as dean of IIRS. He also managed the post of head of the UN-affiliated regional centre with élan. A strong and smiling K. Kalyanaraman was the general manager of aircraft operations and aerial remote sensing.

We had our tasks cut out to take NRSA to the next level of excellence.

I invited the competent, conscientious and trustworthy M.V. Krishna Rao to take over the crucial programme planning and evaluation group, as the existing chief was about to demit office. This group of six became the NRSA management council, whom I leaned on formally and informally. At NRSA, we had a unique confluence of technology, science, application and customer services. It was essential to look at the activities of the five functional entities in this perspective. I formed a system management pool comprising P. Srinivasulu from the data acquisition area, P.V. Narasimha Rao from the applications area and Raghu Venkataraman from the aerial remote sensing team to perform this role as an add-on to their existing engagements.

Along with Krishna Rao, they became my 'brain trust'. I appreciated their ability to have straightforward discussions, where ideas and opinions were exchanged. At times, their ideas would be different from mine. But I encouraged them to air the differing views boldly if they felt these were essential in the overall scheme of things; they did it in the privacy of my cabin. I was fortunate to have such a strong source of constructive alternative opinions that proved vital in many of my decision-making processes.

My next focus was on instilling a spirit of integration among the teams and the partner centres of ISRO as well as on training the younger generation at NRSA so that they could go on to the next level.

In July 2006, I created two forums. The first one ensured formal participation of tier-two functional chiefs for periodic review of programmes and budget; the second one enabled the first three tiers comprising seventy-five scientists and engineers for periodic review of scientific and technical activities on one platform. We devoted a part of the schedule to listen to presentations from the younger scientists and engineers in their late twenties and early thirties. One functionary pointed out that earlier problems were either theirs or ours. Now they were only *our* problems, to be solved collectively.

A 24x7 support centre, equipped to provide space inputs during natural disasters, had become very popular. NRSA had just been chosen by the government to establish and manage a national database for emergency management to enhance disaster preparedness in the country. My recent engagements with the tsunami warning system and close association with the National Disaster Management Authority helped in this new domain.

By late 2006, a new generation Cartosat-2 satellite, capable of imaging up to one-metre spatial resolution, was ready for launch and several ground stations had to be established for this purpose. In simple language, Cartosat-2 was like an agile photographer at an altitude of 630 kilometres that could take a picture of a given small area. The camera was so powerful that even a car or a bridge under construction could be easily identified in the images taken from that height. With this, India joined the club of a few select countries which had come up with similar capabilities. At NRSA, we established ground stations and finished software development for downstream activities of the Cartosat-2 within the stipulated time. Flawless data reception through the Cartosat-2 satellite (launched by PSLV-C7 on 10 January 2007) spelt a moment of pride.

The initial phase of data processing trials and retuning posed many challenges and new learning for both ISRO and NRSA teams.

As the diversity and complexity of the satellites increased, NRSA had to explore technical solutions to cope and be ready to deliver data products soon after each remote sensing satellite was up in orbit.

I believed that the time had come for the NRSA to migrate to ground stations that could be programmed for different missions. Obviously, we needed to adopt a new generation of technologies to minimize the time between satellite imaging and delivery at the user's end. All partner centres of ISRO at managerial and operational levels had to be taken on board. Hence, the 'Integrated Multi-mission Ground Segment for Earth Observation Satellites' (IMGEOS) was conceived.

Re-engineering of the entire chain of data acquisition, processing and dissemination was the starting point. A core group of experts led by deputy director S. Raghunathan and a 100-strong team drawn from NRSA and partner centres of ISRO contributed. The architecture of IMGEOS was ready by December 2006.

One of the outcomes was the physical integration of the entire chain of activities at the Shadnagar complex. My successors, V. Jayaraman and V.K. Dadhwal, implemented it with excellent effect during 2008–11. In November 2011, I came back as the chief of ISRO to participate in the inauguration of this complex and to release satellite data products on the Web within ten minutes of the satellite passing over India.

New Foundations

While at INCOIS, I had realized the importance of developing the institutional campus and its role in facilitating future growth. The main campus of NRSA at Balanagar in Hyderabad was limited for any major expansion. The sprawling 320 acres at Shadnagar housed three ground stations; the skeleton staff posted there was not too happy about travelling to a place eighty kilometres away from the main city. The construction of a new international airport, almost midway between Hyderabad and Shadnagar, was a shot in the arm as I was a campus-lover.

Within two months of taking charge, I had to organize the first meeting of the NRSA Society. I decided to hold it at the Shadnagar complex. Mr Prithviraj Chavan, the minister of state at PMO and also a member of the Space Commission, and Mr Madhavan Nair were impressed by this site located on the Hyderabad–Bengaluru highway and forty minutes away from the airport. I engaged a small team to prepare the master plan for development of the Shadnagar complex, taking into consideration the near-future needs and the long-term development scenario at NRSA. IMGEOS and the national database for emergency management were identified as the core activities for the first phase.

For over a decade, NRSA had functioned as an autonomous body with frozen grants; it had to collect funds for the salaries of its employees and its sustenance. Hence, its primary mandate as the last and crucial link in satellite remote sensing applications was constricted. At times, the imperative national tasks had to take a back seat and NRSA had to dress up as a vendor, bidding for tenders on municipal GIS to enhance its earnings.

A financial restructuring was essential to come out of this situation. I assured DOS/ISRO of enhanced tangible performance from NRSA by upscaling the grant-in-aid. This was one part of the financial restructuring plan. It was accepted and the results were evident in the first year itself; satellite data sales rose by

80 per cent compared to the previous year. NRSA regained its status as a specialist national agency, with the ability to respond to national needs.

We still had a major handicap. The mere label of an autonomous body restricted our access to collateral data and maps from other national agencies. The only option was to restructure NRSA into a government entity like the Vikram Sarabhai Space Centre or the Space Applications Centre under ISRO/DOS.

This was a seminal step and Mr Madhavan Nair found merit in it. We set out to pursue it. The NRSA community, at all levels, could be taken on board as they saw that the decision was in favour of both the organization and its employees. In fact, they were excited with the prospect of having the Ashoka Chakra and the ISRO logo on their identity cards!

The ISRO council welcomed its associate as a member of the family. The governing body and society of NRSA was not averse to the transition as it appreciated the bigger picture. In the entire process, some government functionaries were amused to see that an institution was volunteering to shed its autonomous status when several other institutions were craving for it.

Finally the Space Commission endorsed it in April 2008. After due approvals, NRSA became a government entity from 1 September 2008, rechristened as the National Remote Sensing Centre (NRSC) of ISRO.

India was a founding member of the UN Committee on Peaceful Uses of Outer Space (UN-COPUOS), set up by the UN General Assembly in 1959 to govern the exploration and use of space for the benefit of all humanity—for peace, security and development. The annual sessions of the UN-COPUOS brought all space-faring nations together on a common platform to deliberate and reach consensus on matters of shared concern.

The earlier Indian delegations led by Dr Sarabhai, Prof. Dhawan, Prof. Yash Pal, Prof. Rao, Rajan Sahib, Dr Rangan, Mr Madhavan Nair and Dr Suresh had left imprints in UN-COPUOS from the 1960s. Also, Dr Sarabhai, Prof. Yash Pal and Prof. Rao had shouldered key roles for its three UNISPACE Conferences organized in 1968, 1992 and 1999. This global platform had converged on several initiatives on space applications for the benefit of developing nations. It had begun to focus on the use of space technology for protection of the environment, managing natural resources and to protect the space environment.

In May 2006, the chairman, Mr Madhavan Nair, and the then leader of the Indian delegation, Dr Suresh, zeroed in on me to lead the Indian delegation from the next session (2007) onwards, owing to the impending superannuation of Dr Suresh. I was inducted as a member of the Indian delegation from June 2006. From the IOC of UNESCO to the UN-COPUOS, I had come a long way. Interestingly, my first engagement with the UN-COPUOS in June 2006 was to deliberate on space-based information for disaster management and emergency response. My hands-on experience at INCOIS and later at the decision support centre of NRSA came to my aid to make a lasting impression in my very first appearance at UN-COPUOS.

The expositions on Indian experiences on space applications in diverse domains were star attractions at the annual sessions of the UN-COPUOS and its science and technology subcommittee. I established my credentials as an active contributor through this path.

These helped me to ascend to the position of chair of the Working Group of the S&T subcommittee, an important organ of UN-COPUOS, formed to deliberate on a number of issues such as the use of space technology for socio-economic development and certain organizational matters.

India was to host the 58th International Astronautical Congress from 24–28 September 2007 (IAC 2007). The Hyderabad International Convention Centre was chosen as the venue. I became the chairman of the steering committee for local organizations, working in tandem with Mr Madhavan Nair and Dr Suresh as well as Mr James Zimmerman, the then president of the International Astronautical Federation (IAF).

An annual mega event of world space agencies and space industry, the congress attracted nearly 2500 participants, essentially the who's who of the space fraternity from across the globe. A serial bomb explosion and the Ganesha idol immersion procession at Hyderabad falling on the second day of the Astronautical Congress kept us on our toes; the chief minister, chief secretary and the police department of Andhra Pradesh stood rock steady with us. It was an excellent platform to perform on; NRSA teams at all levels took it as their own family function and put their heart and soul into it.

Besides this mammoth organizational task, NRSA hosted a UN-IAF workshop on space technology for sustainable development, normally held in conjunction with the Astronautical Congress. Prof. Rao, a key architect of the annual UN-IAF workshop held since the early 1990s, was among the audience. He was heard commenting to Dr Suresh, 'Rad did a good job as the lead speaker at the round-table meet of space agencies during the workshop.' At the close of the Astronautical Congress, I received accolades from all possible quarters of the ISRO family, including Prof. Rao and Dr Rangan, for my leadership and organizational skills.

To be honest, I too felt I was doing well in the shoes of director, NRSA, and that I had made a positive impact on the organization on both national and international platforms. Dr Suresh's impending superannuation was triggering a change in the chain of leadership in ISRO. I realized I had positioned myself as a strong candidate with the veracity of my recent contributions. And I was right in guessing so. By the first week of November 2007, Mr Madhavan Nair informed me that I was being considered for director of VSSC.

A few months earlier, Dr Suresh had also hinted that I was being considered for this coveted post, along with a few others from VSSC and ISRO. Dr Suresh had commendably held the post of director of VSSC since September 2003. He was appointed as the director of the newly formed Indian Institute of Space

Science and Technology (IIST) in July 2007. Since then, he had continued to hold additional charge as director, VSSC. The process of identifying a successor to Dr Suresh had been in place ever since his name had gone for IIST.

Rumours around my candidature were on the grapevine as we all assembled at Sriharikota for the launch of PSLV-C8 in April 2007. A few of my old colleagues gave meaningful smiles as I joined Dr Suresh for a prelaunch review there. Soon, there was a meeting of the UN-affiliated regional centre in Kovalam. In the afternoon, I accompanied Mr Madhavan Nair to VSSC to participate in a review of the GSLV project. That added credence to the rumours.

Moreover, I had been an enthusiastic face at Sriharikota during the mission readiness reviews and preparations for the previous four launches. The director of NRSA was expected to be naive in the specialized domain of launch vehicles, and naturally my enthusiasm had amused many young launch vehicle engineers. Two of them, R. Ramavarma and V. Kishorenath, mustered the courage and asked me, 'We heard that you moved from VSSC to the headquarters to NRSA to INCOIS and now you are back at NRSA. What is your next destination?' At that point, I only had a smile to respond with.

The mere possibility of returning to VSSC as its director excited me. I did not want to let this opportunity pass, and I took the responsibility to prepare myself. I did it in earnest. It was a penance of six months. I would regularly wake up at 4 a.m. to brush up my knowledge of launch vehicle technologies and systems and update myself on the recent developments.

I firmly believed that even if I did not make it, my learning would never go to waste.

17

RETURN TO THUMBA

'Sir, I had a small doubt, we are about to create an official profile for you. If you remember your earlier employee code, we can link the profile to your old records. But even if you don't remember it, it does not matter, we'll issue a new employee code and re-enter the records,' the middle-aged administrative officer at VSSC was almost blabbering; he did not want to cut a sorry figure in front of the new director.

But actually he struck the right chord. With juvenile enthusiasm, I exclaimed, 'It's VS25110; you must use the same code.' Elation, emotion, nostalgia, pride—all kinds of feelings bubbled inside me on my first day, after my return, at the Vikram Sarabhai Space Centre, the cradle of 'rocket science' in the country.

One of the walls of the spacious and sophisticated office of the director had photos of all the erstwhile directors—the saintly Dr Brahm Prakash (1972–79), the assertive Dr V.R. Gowariker (1979–85), the meticulous Dr S.C. Gupta (1985–94), the charismatic Mr Pramod P. Kale (1994), the perseverant Dr S. Srinivasan (1994–99), the go-getter Mr G. Madhavan Nair (1999–2003) and of course my enduring friend, philosopher and guide Dr Suresh (since 2003) from whom I was to take over this great institution.

It was quite a big bite to swallow in one go as almost all my predecessors had made indelible contributions to the Indian space programme. I was determined to keep up the great tradition and prove to be a worthy successor.

VSSC, as we all know, is the largest centre of ISRO and thanks to the diversity in the nature of research and development, and the profile of scientists and engineers working here, it is often referred to as the centre of centres within ISRO. Among the illustrious alumnus of VSSC (other than the directors whom I just mentioned) there were Kalam Sir, Kurup Sir, Dr Muthunayagam, Mr D. Easwara Das and Mr Aravamudan, who were also the founders of the Indian space programme. India's first launcher, SLV-3, took shape here, and over the last forty years the next generation of launchers, ASLV, PSLV, GSLV and LVM3, came up. It was wonderful to come back to where it all began.

On a bright morning on 30 November 2007, I could sense that we had to navigate through a delicate transition process. Both Dr Suresh and I were protégés of Dr Gupta and hence we decided to start the day by visiting him at his house, to seek his blessings for our respective careers. The formal transition took place in a simple function organized in the director's office. The entire VSSC community assembled in the large open foyer and paid tributes to their outgoing director; they welcomed his successor, the new face from Hyderabad.

During the ceremony, the lively presence of P.S. Veeraraghavan, the director of the ISRO Inertial Systems Unit, was noticed by one and all. He was one of the most senior scientists in the VSSC system. By then, a congratulatory message came in from Dr T.K. Alex, the then director of ISRO's Laboratory for Electro-Optics Systems (LEOS)—a specialized unit that developed the eyes and ears of ISRO's satellites. I was deeply moved and humbled by their gracious behaviour; they had stood with me during trying times at Thiruvananthapuram and later at Bengaluru.

Earlier that day, Dr Suresh chaired his farewell meeting and I was introduced to the VSSC Centre Management Council that had, other than the director, fourteen top executives under its wing, representing different technical domains. Most of them were known to me since my days at VSSC and the headquarters. The council comprised eight deputy directors, the chief controller of administration, project directors of three main projects (PSLV, GSLV and LVM3), the associate director and director (projects).

Along with my elevation as director, some parallel adjustment was also done at VSSC. S. Ramakrishnan, associate director (projects), was elevated to director (projects). Adimurthy, a brilliant aeronautics expert, agreed to continue as associate director (research and development) during his extended service. Other than the eight deputy directors heading diverse technology domains, there were N. Narayana Moorthy as project director of LVM3, G. Ravindranath as project director of GSLV and George Koshy as the project director of PSLV. Most of the deputy directors (Selvaraju, Chidambaram, Bijan Das, John Zachaiah and Enamuthu) were my contemporaries. Two of them, K.N. Ninan and P.P. Sinha, both on extended service, had been my seniors during the 1970s. T.S. Ramadevi (who wished me luck on the day of my recruitment in Thumba in 1971) was my teacher in college!

When my turn came to speak at the farewell party, I just recollected anecdotes from my past interactions with each one of them and shared my admiration for them. As we all dispersed, a bond of mutual respect and warmth was established. This was succinctly summarized by Adimurthy, who presented the vote of thanks.

After handing over the charge to me, Dr Suresh shared a few words of wisdom and advice. The concluding line of his speech left a deep impact on me: 'VSSC is such a robust organization; anyone will have to struggle really hard to spoil it.' Serious advice was hidden in these words.

Soon Dr Suresh moved to his temporary office that was set up in the adjacent room for his new role as the founder-director of the Indian Institute of Space Science and Technology. The epitome of dignity, he demonstrated traits of an ideal predecessor, leaving me to do my job but making himself available whenever I needed him. He had the assurance of unstinted support from VSSC to fulfil his new responsibilities. That was the most memorable and delightful transition of my career.

A Huge Responsibility

The exchange of pleasantries was over, and we soon got down to business. For me, this was a challenging shift. To lead this centre of centres, I had to gain insights into an assortment of technologies to bridge the gap of two and a half decades. Administratively too, in terms of managing human capital, it was quite huge and very different from the set-up at NRSA and my baby, INCOIS.

But the cardinal change in VSSC, from late 1980s to the present, was in the leap of multidisciplinary technology development. I remembered and engraved in myself that I had been placed on a prestigious platform, and I had to perform.

The first month was essentially a phase of familiarization, adaptation and gaining acceptance on the new turf. Nearly 80 per cent of human capital at VSSC had been inducted after my exit in 1981. Frankly, I went through the bulky telephone directory of VSSC to familiarize myself with the key individuals in each discipline. I had observed many of them as I had been part of their assessment interviews since 1992 at ISRO. But we needed to know each other. About 25 per cent across the board at all levels were youngsters in their twenties and early thirties. I needed to connect with this new generation.

I spent a considerable amount of time on each of the technical entities and projects, meeting the key persons, mingling with the team members and moving around facilities and workshops. I listened to them to understand their activities and gave glimpses of my relevant past associations. My study of space system engineering and space transportation systems over the past six months, and thorough preparation before each of these sessions helped me through this.

I got down to the brass tacks of rocket science. My familiarity with PSLV in my previous stint at VSSC and the headquarters gave me the much-needed kick-start.

PSLV, a four-stage launcher with six solid rockets strapped around the first stage, was conceived in 1978. Its purpose was to place a remote sensing satellite of 600 kilograms into a polar sun-synchronous orbit of 500–1000 kilometres around the earth. Our optical remote sensing satellites could capture images of a certain area over a certain period of time, only if they followed an orbit that went around the North and South Poles longitudinally. This was called the 'polar' orbit. A stringent requirement for the orbit was to maintain a constant angle to the sun–earth line. This ensured imaging of any area at a constant local time or light conditions. As PSLV matured over the years, the power and versatility too were enhanced. PSLV was utilized for launching several satellites into different orbits around the earth, including geostationary satellites.

PSLV is an intricate marvel of system engineering. During the eighteen minutes of its nominal flight, nearly forty rockets (of either solid or liquid propellants) power its ascent against gravity, separate the spent stages and guide it through its trajectory. These rockets have to perform within definite timelines and detach on their own after their tasks are through. PSLV navigates and guides itself with precise sensors and complex computers at every stage. It is robust enough to endure the perturbations encountered en route its flight. A few hundred interconnected electronic packages and fluid control components have to function flawlessly for a successful flight. This is essentially the design part of the story.

But, before a mission, the PSLV gets assembled at the launch site; detailed analysis, modelling and simulation as well as rigorous quality control are ensured for its success. Non-conformance or a flaw in the functioning of even the most miniscule element, can mar the mission. And a launch failure is a shameful spectacle as national pride is associated with it. This was the tough part of being associated with the launch vehicle domain.

I had to prepare myself for the new role. I leaned on experts of launcher technology, K. Sivan and S. Somanath, to learn the nuances of the PSLV launcher and expectations from its multiple elements during a flight.

The intense phase of my learning continued for the next two years. Somanath was my main tutor and, over time, he evolved as my conscience keeper and conscientious critic. I learnt about the advances in aeronautics and cryogenic propulsion from two admirable youngsters, V. Ashok and B. Deependran. The list goes on.

I had picked up fifteen brilliant youngsters and made a pool of scientific staff officers. This was an add-on to their existing roles, but each one them had to give analytical inputs on domains beyond their current engagement. Mostly, we sat together on weekends. They too benefited as their domain knowledge expanded, and they were able to prepare themselves for bigger roles.

I also realized that in a huge multidisciplinary set-up like VSSC, where there were thousands of brilliant and motivated engineers, the job of the leader went beyond that of a technology mastermind. Being an expert did help but what mattered more was to bring the entire population on to the same thinking plane. I strongly believed that my decisions on technical and organizational matters had to be a function of collective knowledge and experience of the men and women I was leading. I felt that clarity on organizational objectives and priorities had to be set right away. The leader has the mandate to communicate his priorities and at the same time provide an environment to achieve them. I tried my best to follow this mantra.

My first probationary test came rather quickly, within forty days of taking over. I had to oversee the launch of PSLV-C10, for which the mission campaign had almost reached the final phase. I felt that Dr Suresh deserved the lion's share of the credit for this launch.

The launch took place on 21 January 2008. I occupied the hot seat of director, VSSC, at the Mission Control Centre in Sriharikota, next to the chairman of ISRO. So far, my acquaintance with the Mission Control Centre was either as a staff officer to the chairman or most recently as an observing director from NRSA. But this time, I made the most of my recent learnings on PSLV by sitting in the best possible place.

The next launch of PSLV (PSLV-C9) was getting ready to orbit the Cartosat-2A, IMS-1, an Indian mini satellite, and eight small foreign satellites. Incidentally, this mission was unique for me. As director of VSSC, I was responsible for the PSLV-C9 launcher that would deliver the satellites in their desired orbits, and at the same time I was responsible for the last chain of the mission, i.e. to receive data from Cartosat-2 at NRSA. I was additionally the director of NRSA at that time.

PSLV-C9 was unique in another aspect, it was to launch ten satellites at one go (which we had never done before), and the payload dispensing technique had to be modified to ensure that these satellites did not get in each other's way. A new guidance software, which was introduced for the PSLV-C9 mission, came in the critical path of the launch schedule and I had to wade in. U.P. Rajiv helped me update my basics of the mission software and guidance algorithm that I had learnt in the 1970s. He tutored me on both the physical process and the mathematical formulation. This helped me in reviewing the guidance design and simulations with my friend Bijan Das's teams that cracked the problem the next month.

Most importantly, we had a successful launch of ten satellites by PSLV-C9 on 28 April 2008. We set a record for launching the largest number of satellites in a single mission. That launch caught a lot of public attention.

After the two successful missions under my charge, I concentrated on organizational development.

Shaping the Future

VSSC, the largest centre of ISRO, always had a very distinct culture. Since the 1980s, there had been decentralization and democracy in decision-making, delegated to the chiefs of entities and projects. More importantly, there was a culture of openness, transparency, trust and tolerance to internal criticism. Preserving and enhancing this ambience was my foremost priority.

While S. Ramakrishnan, as director (projects), was asked to focus more on execution of projects, Adimurthy was tasked with shouldering the oversight of research and development at the centre. Together, we formed a trinity at the head table of the centre's management. We co-chaired the Space Scientific Committee comprising 225 senior functionaries that met on the first Friday of every month with a structured agenda of setting targets and monitoring accomplishments. We also initiated a slot dedicated for younger members (those in their mid-career) to make presentations on a technical area of their choice. Everyone strived not to miss it. Job rotation was accepted as a path to success by these future leaders.

I prioritized attending all technical, managerial and administrative issues over everything else. I believed in leading from the front, and involved functionaries to resolve issues at hand. What I enjoyed during my two-year tenure at VSSC was the way such problems were brought forward and the collective process adopted to resolve them.

A classic example was the induction of a new mission computer built around a 16-bit processor, Vikram 1601, for PSLV and GSLV launchers. At the equipment level, it had gone through a lengthy process of tests, simulations and flight trials a year earlier. It was slated to be inducted as the prime chain of computing in PSLV for its navigation, guidance and control functions, while its forerunner, with an INTEL i960 processor, would continue as the hot standby. Both the prime chain and standby chain had to be in harmony throughout the flight, under all possible conditions.

During one of the rehearsals of the automatic launch sequence at Sriharikota, we observed an anomaly. P.S. Veeraraghavan, T.R. Chidambaram and Santhakumari were quick to foresee its gravity. Nothing could be left to chance in a space mission. I set up a 'tiger team' of thirty software professionals, led by the chief of software reliability, Vikraman Nair, for a zero-based software audit and to unravel the remnant bugs in the flight software. A series of simulations and launch rehearsals followed. After all formal reviews, we thought it was important to get the final judgement from Dr Gupta, the founding father of our launch vehicle avionics and flight computing. As I gave a detailed overview of the symptoms, diagnosis, corrections and confirmations to Dr Gupta in the presence of the 'tiger team', his smile confirmed that I had grasped the problem. At the end of the long session, he said, 'Your team has the right understanding of the problem. I am confident that their solution will work.'

Once, we had to go through the trauma of a major accident at a construction site in Thumba; thirty-five construction labourers were on the site at the time of the accident. A high-rise slab had collapsed while work was in progress. Some of the labourers had got trapped under the debris. In a politically sensitive state like Kerala, an accident of this order on the premises of a Central government agency could have caused major unrest. This had the potential of snowballing into a much bigger event, if not reported on time. I called up the chairman, Mr Madhavan Nair, and requested him to alert Delhi. Within ten minutes of the episode, I reached the accident site and stood by the rescue operations that went on till midnight. A sizeable team consisting of engineers and members of service organizations of VSSC complemented the efforts of the Central Industrial Security Force (CISF). The district collector, the state police department and local hospitals stood by us and provided medical attention to all the victims. The casualty list was contained to only two. We felt it was our duty to brief the media who had been patient even though they were not allowed inside to cover the accident.

Engaging with the younger generation in VSSC is something I cherish the most. VSSC had a vibrant pool of about a 1000 engineers, below the age of thirty-five. This generation had, on an average, spent about six to eight years in the organization. These formative years of their career eventually determined their career choices and through them the future of the organization too. Ramakrishnan, Adimurthy and I decided to spend some quality time with them. We made it a practice to interact with them in the director's conference hall. The youngsters would be called in batches of twenty; groups were made based on their year of induction. The weekly interaction would typically last an hour, where they would briefly describe their work. The three of us would follow it up with a few questions, and then we would ask them to open up on their aspirations, apprehensions and doubts. These were freewheeling discussions and we got some fantastic insights into the psyche of our future assets. The youngsters found it a rewarding experience as they were benchmarked against their own batchmates spread across domains and roles.

I remember one very impressive youngster, Bhuyan, who was in the GSLV project management team. He did not like the idea of frequently travelling from Thiruvananthapuram (VSSC) to Mahendragiri (Liquid Propulsion Systems Centre or LPSC) to monitor the assembling and testing of the cryogenic stage. I tried to encourage him to look at that role a bit differently. 'Do you realize that for the last sixteen years ISRO has been toiling hard to make cryogenics work for GSLV. The work you are doing has national importance attached to it. Long after we succeed, long after you retire, you will recollect this drudgery as your contribution to something great that the country achieved,' I said. He appeared convinced, but his actions spoke louder than his words. The transformation was evident when he came up with a novel test methodology to attempt high-altitude testing of the Indian cryogenic engine of GSLV, at Mahendragiri. ISRO recognizes these amazing young guns. I was happy to see his name as one of the awardees of the Young Scientist Merit award.

By April 2008, after thirteen launches and twelve consecutive successes, PSLV had reached maturity. The Chandrayaan-1 mission was the cynosure of all eyes in ISRO, and, of course, it was a national imperative which we had to deliver perfectly.

During the mission design, it was discovered that the existing capability of PSLV was not enough to take the Indian dream to the lunar neighbourhood. We had to develop a high-end version of the PSLV (the XL version).

The then President of India, Kalam Sir, wanted an India-made object to touch the lunar surface. Hence, the Moon Impact Probe payload was conceived and it resulted in extra mass for the Chandrayaan-1. After detailed design and analyses, the teams at VSSC proposed to upgrade the standard strap-on solid motors. In the generic PSLV, the core solid motor (S139, a solid 139-tonne motor) was bolstered with six small solid rockets, each with nine-tonne propellants strapped around it. In the new scheme we proposed to use six units of twelve-tonne strap-on motors.

This decision had three major impacts—on the lift-off dynamics, aerodynamic characteristics, and controllability in the early phase of ascent. A series of wind tunnel tests, computational fluid dynamics modelling and confirmatory analysis had to be carried out by VSSC teams before inducting this upgraded version of PSLV (designated as the PSLV-XL series with a lift-off mass of 320 tonnes) as the PSLV-C11 was to launch Chandrayaan-1.

The Moon Impact Probe was also a challenge as we had to ensure a smooth separation from an orbiting Chandrayaan-1 spacecraft as well as a metal-free solid propulsion system for spin and de-boost of the probe in the lunar environment. This was a fantastic challenge and nothing motivated the VSSC engineers more. The entire centre rose to the occasion. We promised ourselves to give our best and the preliminary results appeared encouraging. With a lot of enthusiasm and determination, we started the launch campaign in Sriharikota. It was to be a memorable experience for the country.

While I kept myself busy in helping VSSC contribute in one of the biggest technology projects of the country, another interesting development was taking place in the Department of Ocean Development. The department had been elevated to the Ministry of Earth Sciences (MoES). Dr P.S. Goel, the then secretary, superannuated at the end of April 2008 and a process for the selection of his successor was put in place by the government. The search committee was keen to consider my candidature and I was approached by a few known quarters. It would have been inappropriate and unethical for me to go back to MoES for personal gains. I decided to seek frank advice from my fatherly figure, Prof. Rao.

Fortunately, Prof. Rao and his wife came to Thiruvananthapuram for a couple of days in the first week of June 2008. I joined them for breakfast at the Hotel Taj Samudra in Kovalam. We talked and I revealed my intention of not leaving ISRO for any other opportunity. But Prof. Rao cautioned me and said, 'Rad, look at your age, you will be sixty in a year's time. You might just miss the bus if you wait for such a role in ISRO.' Finally, I told him, 'Sir, in the worst scenario, I might get stagnated and superannuate as director of VSSC; that is fine with me. I don't want to be an opportunist.' He wore that beautiful smile of his and said, 'Rad, I agree with you . . .'

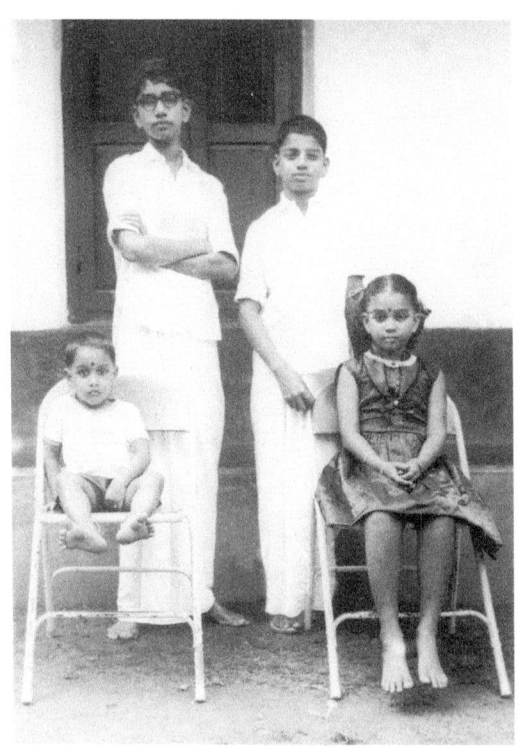

With my siblings, Sivadasan (standing on the left) and Radha, and cousin Krishnakumar (seated on the left) in our ancestral house, 'Koppillil', in 1964, right in front of the room where I was born fifteen years earlier.

All geared up for a postgraduate programme in management at IIM Bangalore in 1974.

The bridegroom with his parents in front of 'Koppillil' on 1 July 1983.

With my wife, Mini, in the sprawling precincts of Koodalmanikyam Temple,
soon after our wedding ceremony on 1 July 1983.

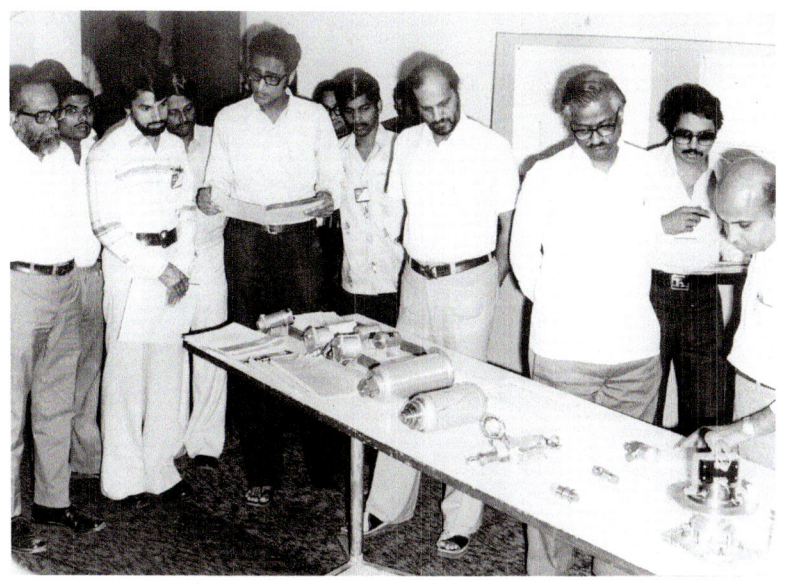

With my mentor, the director of the Avionics Group, Dr S.C. Gupta,
during a briefing session at VSSC in February 1980.
(From left to right) Dr S. Ramnath, me, Mr R. Aravamudan,
Dr V.R. Gowariker (director, VSSC), Dr Raja Ramanna and Dr S.C. Gupta.

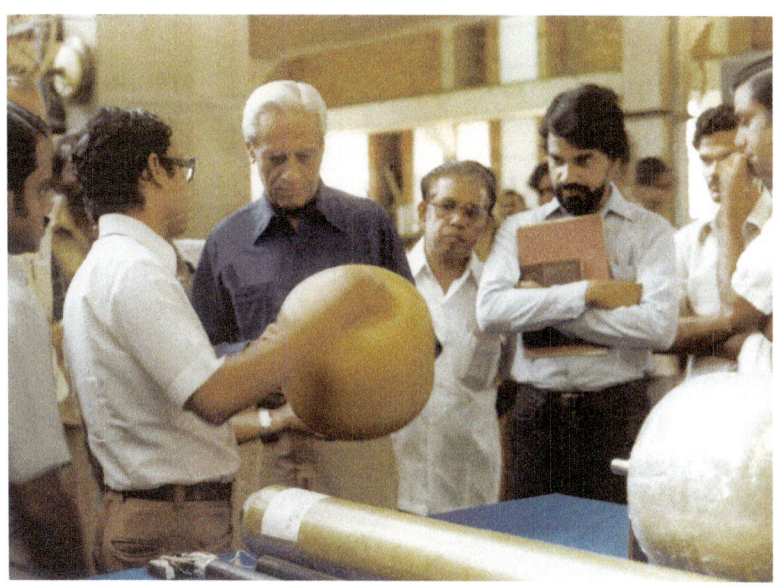

Accompanying the chairman, Prof. Satish Dhawan, during a visit to a
technical facility at VSSC in September 1984.

Accompanying the chairman, Prof. U.R. Rao, during Prime Minister Rajiv Gandhi's visit to witness the launch of ASLV-D1 on 24 March 1987. (From left to right) Mr Gandhi, the chief minister of Andhra Pradesh, N.T. Rama Rao, me and Prof. Rao.

A review session of the GSLV-D3 held in the first week of April 2010 at the launch pad in Sriharikota. Flanked by Chandradathan and M.Y.S. Prasad on my right and left; Ravindranath and Mohammad Muslim are standing behind.

With my predecessors prior to the launch of GSLV-D3 on 15 April 2010.
(From left to right) Prof. U.R. Rao, Mr G. Madhavan Nair, me and
Dr K. Kasturirangan.

A leader must take responsibility for his/her failures: Facing the national media
after the failure of the GSLV-D3 on 15 April 2010.

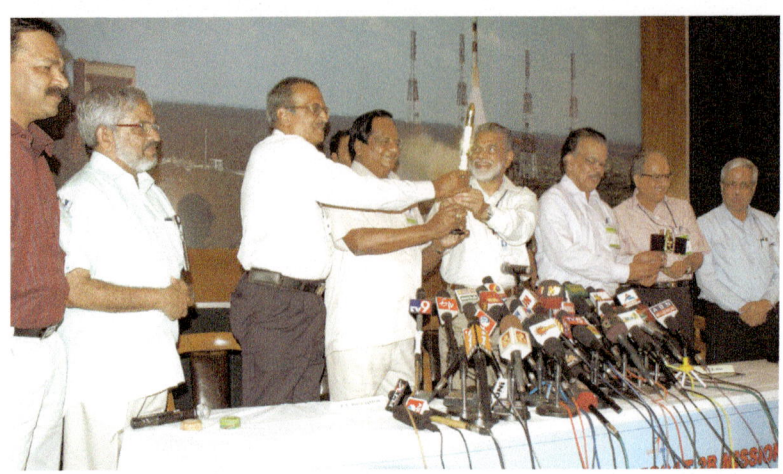

Credit multiplies when shared with the team: The joyous Team ISRO after the successful launch of the PSLV-C15/Cartosat-2B mission.
(From left to right) The project director of PSLV-C15, P. Kunhikrishnan, centre directors, M. Chandradathan, S. Ramakrishnan and P.S. Veeraraghavan, me, centre directors, T.K. Alex, R.R. Navalgund, V. Jayaraman, and the project director of Cartosat-2B, M. Krishnaswamy (out of the frame).

Initiating cooperation with NASA on earth observation, planetary exploration and space education. With NASA administrator Major Charles F. Bolden Jr at the International Astronautical Congress, held at Prague in September 2010.
(From left to right) Somanath, Dr B.N. Suresh, V.S. Hegde, Surinder Singh, the late P.M. Bala (standing behind Surinder Singh), Nilanjan Routh, Shantanu Choudhuri, V.K. Dadhwal, Prof. J.N. Goswami, all members of the Indian delegation.

I again chose to face the national media alone after the failure of the
GSLV-F06 on 25 December 2010.

On 26 March 2011, Prime Minister Dr Manmohan Singh, in a wonderful
gesture of solidarity, visited the Space Application Centre in Ahmedabad
during ISRO's difficult times, and mingled with the scientists and engineers.
The then chief minister of Gujarat, Mr Narendra Modi,
was the guest of honour.

In the 'hot seat' at the Mission Control Centre during the launch of the PSLV-C19, carrying India's first microwave satellite, RISAT-1, on 26 April 2012.

Prime Minister Dr Manmohan Singh congratulates me before boarding the helicopter at Sriharikota after the successful launch of PSLV-C21. The PM had come to attend the 100th mission of ISRO.

President Pranab Mukherjee at the Mission Control Centre, Sriharikota, on 25 February 2013, after PSLV-C20 orbited the Indo-French satellite, SARAL, and six more foreign satellites.
(From left to right) Jose Antonio Sanz (project manager, MDA Systems, Canada), Sir Martin Sweeting (chairman, Surrey Satellite Technology Ltd, UK), me and Mr Yannick d'Escatha (president, French National Space Agency).

India's first navigation satellite, IRNSS-1A, atop PSLV-C22, awaiting its ferry to the outer space. The launch took place on 1 July 2013.

En route to the red planet: 'Mangalyaan' (MOM) commences its voyage from earth to Mars on 5 November 2013, aboard the PSLV-C25.

Sharing rare moments after PSLV-C25 placed 'Mangalyaan' (MOM) into an elliptical orbit around earth, the first crucial milestone in its voyage. (From left to right) Dr K. Kasturirangan, US Ambassador Nancy Jo Powell, Union Minister V. Narayanasamy, Mr Pulok Chatterjee (principal secretary to the PM), Prof. U.R. Rao and other senior scientists and invitees.

A success achieved after many failures: The lift-off of the GSLV-D5 with the Indian cryogenic upper stage, on course for the landmark victory of ISRO, on 5 January 2014.

Presenting a souvenir of the GSLV-D5 lift-off to Prime Minister Dr Manmohan Singh on 21 January 2014 at '7RCR', New Delhi. The project director, Dr K. Sivan, looks on.

Receiving the Padma Bhushan award from President Pranab Mukherjee on 26 April 2014.

Introducing my senior colleagues to Prime Minister Narendra Modi after the successful launch of PSLC-C23 on 30 June 2014 at the Mission Control Centre in Sriharikota.
(From left to right) Kunhikrishnan (project director, PSLV),
S.K. Shivakumar (director, ISAC), me, Prime Minister Narendra Modi,
Union Minister Venkaiah Naidu and the chief minister of Andhra Pradesh,
N. Chandrababu Naidu.

An anxious wait of ten minutes to confirm whether the liquid engine of 'Mangalyaan' (MOM) obeyed the command to restart after a hiatus of 296 days.
(From left to right) Pichamany of ISTRAC, S.K. Shivakumar, Dr K. Kasturirangan, me, Prof. U.R. Rao, A.S. Kiran Kumar and V. Koteswara Rao at ISRO's Mission Operations Complex in Bengaluru.

Aaj Mangal ko MOM mil gayee: The historic announcement by Prime Minister Narendra Modi on 24 September 2014. The mission director, Kesava Raju, is on the extreme right with his mission team.

A perfect maiden lift-off of India's future launcher LVM3
on 18 December 2014.

My address after the success of the LVM3 experimental flight. I was flanked
by project directors Somanath (on my right) and Unnikrishnan Nair (on my
left). My beloved team members are standing right behind.

A Kathakali performance in Bengaluru in January 1987.
I played the role of King Daksha.

A Kathakali performance in December 1995 in Bengaluru. I played a
Brahmin father, who is pleading with Lord Krishna because his nine children
died immediately after birth.

Singing *Sree Krishnam Bhaja Maanasa Sathatham Srithajana Paripalams* (mind, chant the name of Sree Krishna always; he protects those who seek Him) during a Carnatic music stage performance.

Our family got together for Dhanya's wedding on 26 December 2011 in Bengaluru. Mini and I are seated with my aunt, Kallianikutty Amma, (in the centre), our siblings and their spouses, children and close relatives.

18

THE LIFT-OFF

'10 . . . 9 . . . 8 . . . 7 . . . 6 . . . 5 . . . 4 . . . 3 . . . 2 . . . 1 . . . Lift-off normal,' announced the range operations director on the internal address system.

We breathed a great sigh of relief—Chandrayaan-1, the first Indian lunar mission, was on its glorious trajectory towards the moon atop our newly upgraded PSLV. The launch of Chandrayaan-1 is etched in the memory of every Indian; it was probably one of the most prolific technological milestones that we had achieved until that time; 22 October 2008 was a very very special day for all of us.

The build-up to this historic day was a bit shaky towards the end. The forty-nine-hour-long countdown process, in which the vehicle goes through several checks and fueling operations, progressed smoothly till the last twelve hours. And then there was a hostile spell of torrential rain and severe thunderstorms of a typical north-eastern monsoon.

To make things worse, the previous afternoon, a leak was diagnosed at the tricky tibia joint between the propellant filling unit and the launcher. There was a flurry of activity overnight. We had to be fully alert as we were handling a hypergolic fuel–oxidizer combination. This meant spontaneous ignition in case the two came into contact. Probable simultaneous leaks in both the fuel and oxidizer lines could have caused a catastrophic fire. As such, the torrential rain was a blessing in this context. Probably the rain gods were also on our side to pull it off.

The tenacity and alacrity of a few hundred ISRO engineers, assembled at Sriharikota, were put to test. The propellant filling was restored but with severe constraints in its rate of flow and its ideal ratio of the fuel UH 25 (a mix of unsymmetrical dimethylhydrazine and hydrazine hydrate) and the oxidizer (nitrogen tetroxide). We had lost a few precious hours earmarked for the ensuing countdown operations. The director of projects, Ramakrishnan, and the range operations director, M.Y.S. Prasad, came forward with contingency plans. Sivan's mission team (he was the group director of the Guidance and Mission Simulation at VSSC) predicted the possible fallout and guaranteed that adequate margins existed for a successful mission. Our captain, Mr Madhavan Nair, was concerned but kept his composure throughout as the anchorman. He finally took the call and gave us the go ahead for the launch. It was a tough call on his part.

The rain receded and the moon shone bright, inviting us to take on the challenge. Finally, PSLV-C11 was ready for lift-off at 0622 hours. This lift-off time had a stiff requirement, dictated by the earth-moon geometry of that day.

The entire nation and even many from abroad were glued to their television sets as the mission director, George Koshy, authorized the automatic launch sequence computer to take over the last ten minutes of operations. They keenly observed the body language and facial expressions of the mission executives as PSLV started its ascent.

Within eighteen minutes, the PSLV-C11 did its assigned job and provided the Chandrayaan-1 spacecraft enough velocity to put itself precisely into an elliptical orbit around the earth for its voyage towards the moon over the next three weeks.

As the announcement of the successful injection came in, Mr Madhavan Nair spontaneously hugged me with great jubilance and warmth. He had held the post of director, VSSC, and was the project director of PSLV for its first two flights; he knew the challenges we had faced. We, the ISRO council members, proudly stood behind him as he announced the success of the PSLV-C11 mission on national television.

That was the dawn of a new era for the Indian space programme. Mr Madhavan Nair was rightfully the hero of the mission, and was aptly titled the 'Moon Man'. Being his close associate in my capacity as the director of VSSC, I rocketed to national fame and became one of the front-line public communicators from ISRO. More importantly, I became the pride of my home town as well as the schools and colleges where I had studied.

The Next Big Leap

ISRO learnt several important lessons from Chandrayaan-1. The limited capacity available with the upgraded PSLV forced us to deduce a smart mission strategy to reach an orbit near the moon. This was accomplished by a series of orbit-raising operations around the earth by using the propulsion system of the spacecraft itself. Also, we demonstrated the restart capability of the liquid engine in the spacecraft after a fortnight, for the lunar insertion manoeuvre. Earlier, we had built this complex spacecraft with eleven scientific instruments from India and abroad, managing several protracted international interfaces. We learnt and put into practice the provisions of protecting the electronics systems from high electromagnetic radiations and the harsh thermal environment in lunar orbits. We tested our own precise autonomous navigation system for launchers and spacecraft. We completed a deep space mission, covering a range of nearly 4,00,000 kilometres, in the very first attempt.

All this knowledge came in handy for our Mars Mission that took place six years later. By all means Chandrayaan-1 was the springboard for ISRO and personally for me, as the experience was useful at a later stage.

The guiding vision of the Indian space programme articulated by Dr Vikram Sarabhai in the late 1960s states thus:

There are some who question the relevance of space activities in a developing nation. To us, there is no ambiguity of purpose. We do not have the fantasy of competing with the economically advanced nations in the exploration of moon or the other planets or manned spaceflight. But we are convinced that if we are to play a meaningful role nationally and in the comity of nations, we must be second to none in the application of advanced technologies to the problems of man and society, which we find in our country. . . .

But, over the decades, India's capabilities in the frontier of outer space did grow significantly. Our developmental imperatives, strategic perspectives and aspirations did expand. India stepped into exploration of the moon three decades after the vision was stated, without diverting from the cardinal intent of the founder.

India occupied the front table amidst the comity of space-faring nations with primacy in space applications and praiseworthy records in satellites and launchers. However, it was yet to open its account in human spaceflight although the famed Indian Air Force pilot, wing commander Rakesh Sharma, had flown aboard a Soyuz T-11 for the USSR in 1984. ISRO had also executed a successful space capsule recovery experiment in January 2007.

Traditionally, Russia and the USA had had a long heritage in human spaceflights since the early 1960s. China too entered this arena and demonstrated their capabilities in 2008. Europe and Japan became partners of the US-initiated international space station programme. All these spacefaring majors had a long-term vision of human presence on the moon by 2020 and later on Mars.

Honestly, till 2006 our focus was on a citizen-centric space programme, and we never really gave a serious thought to development of technologies for such advanced missions.

A human spaceflight programme would have been extensive in terms of development budgets and it did not have a near-term impact on our focused citizen-centric space programme. But inquisitiveness kept knocking on our doors and the results were the Space Capsule Recovery Experiment, Chandrayaan-1, Mars Mission, ASTROSAT and so on. Mr Madhavan Nair, the adventurer that he was, pushed for a human spaceflight programme in 2006. His ideas were ably supported by Dr Suresh, the then director of the Vikram Sarabhai Space Centre, who also assembled a study team.

I had been aware of these developments in ISRO as director of NRSA and even earlier. As a member of the ISRO family, I had a firm belief that ISRO was in a ripe state to establish a human presence in the solar system along with the front runners. I felt that the human spaceflight programme should be a prime portfolio of the Indian space programme along with a sensible mix of space robotics.

I inherited this fantastic proposition when I took over VSSC and vowed to take it forward. The Space Commission had just approved a pre-project investment for new critical technologies. Our deliverables were pretty clear. VSSC was mandated to come up with a programme for development of an indigenous space capsule and a reliable launch vehicle capable of carrying a two-member crew to a low earth orbit of an altitude of 275–400 kilometres, keep them for a week in space and bring them back safely to land in the Bay of Bengal or the Arabian Sea. Much of the spadework for the project report had already been done before the baton was handed over to me.

On my first working day at VSSC, I telephoned Unnikrishnan Nair, convener of the feasibility study team and asked, 'Unnikrishnan, how busy are you this afternoon? I wish to meet you.' I had observed his enthusiasm while assisting his team leader, Madan Lal, for a sterling defence of the feasibility study, in November 2006, to a national group of eminent engineers and scientists assembled at Antariksh Bhavan. This was the first technical review that I took in VSSC that afternoon.

The idea of a human spaceflight kindled the imagination of the entire ISRO family. Mr Madhavan Nair and I set up a vibrant inter-centre team of about 100 scientists and engineers, who were young enough to see its logical conclusion. We chose Unnikrishnan Nair as the project director, who later became the youngest project director for the largest project in ISRO. S. Ramakrishnan was assigned an additional role of chief executive; his job was to mentor this young and ebullient team. As the director of VSSC, I chaired the project management council that had the concerned centre directors on board.

We knew the task at hand was arduous.

When we discussed the crew capsule, we realized that there were many important systems that had to be developed. The environmental control and life support systems, the crew escape systems—these were unchartered domains of technology for us. We had to develop a launcher, which would be capable of taking the 6000-kilogram crew module safely into the low earth orbit. While our GSLV or LVM3 had the energy to do so, these launchers had to achieve the specified reliability goal of 0.98, i.e. two failures in 100 flights.

Our GSLV had had only five flights with two successes and LVM3 was nowhere near the finishing line. Obviously, we had to prove that these launchers were fit to orbit our (unmanned) spacecraft, and later carry out a host of enhancements in the design, manufacturing and rigorous testing to complete a process called 'human-rating'. With a long-term perspective, although not in our deliverable list, we needed to set up a few facilities such as an astronaut training centre, provisions at the launch pad for handling astronauts, and enhancements at mission control centres to communicate with astronauts while they were orbiting around earth.

I was excited at the opportunity and immersed myself in the project with the tiny yet sparkling core team of Unnikrishnan, Sunil, Vinod, Mohankumar, Harish, Anand, Kumar and Ganju. We studied and learnt together and thus geared ourselves up to enlighten the decision makers in Delhi. Obviously we could not just throw jargon at them; we had to orient ourselves to a language which was comprehensible.

True to ISRO's traditions, we set near impossible targets of executing the first human spaceflight within seven years from the formal kick-off. Prof. Rao promised to deliver the first Indian satellite within three years when we did not even have a space standard laboratory. He did deliver it in three and a half years. We projected a financial outlay of Rs 12,400 crore (at 2008 prices) for this initiative to be accomplished by ISRO with several national agencies and industries as a national programme. We waltzed through the review of the Draft Project Report by the National Committee of Eminent Engineers and Scientists that took place in February 2008.

Meanwhile, in a bilateral meeting, Mr Madhavan Nair and the chief of the Russian Federal Space Agency, Anatoly Perminov, decided to explore possible areas of cooperation in human spaceflight. I was the co-chair, representing India in the Joint Working Group that came up with a protocol that suggested 'a flight of two-member Indian crew in a Soyuz Spacecraft by 2012–13 along with flight testing of Indian avionics in Soyuz as a forerunner to the human spaceflight using Indian Launch Vehicle and crew module'. Two feasibility studies were also mounted with the Russians on commercial terms.

Later, I co-chaired a joint working group with NASA to discuss their involvement in the human spaceflight programme. Of course there were technological challenges that we could solve ourselves, and there were technologies which were available only through international cooperation. We strongly felt that in view of the target schedule there was no need to reinvent the wheel.

I was asked to make a presentation on the project at the 110th meeting of the Space Commission. Incidentally, the meeting happened on 12 April 2008, the forty-seventh anniversary of the first human spaceflight by the Soviet Union. As I made the presentation, there was a visible flow of sentiments with the latent euphoria of attempting something similar in our own country. We established our case before the Space Commission who assigned us more homework to confirm whether this new initiative would in any way hamper the current national obligations of ISRO. Finally, in August 2008, the Space Commission endorsed our project report to be placed before the Union cabinet; this was the last hurdle.

As a normal practice, our project report was sent for inter-ministerial consultations as a prelude to placing it before the cabinet. In the first week of February 2008, the prime minister's office mandated a seven-member national committee to examine the 'desirability and need' of the Indian human spaceflight mission. I was privileged to be chosen as a member along with Mr Montek Singh Ahluwalia (the deputy chairman of the Planning Commission), Dr R. Chidambaram (principal scientific adviser to the Government of India), Prof. Roddam Narasimha, Prof. M.G.K Menon, Prof. Yash Pal and Dr M.S. Swaminathan.

This committee addressed and deliberated on the global scenario, imperatives, technological capability in the relevant domains, choice of launcher and its heritage, risk assessment, crew safety and mission assurance, possible collaboration with other space agencies, the role of the Indian space industry and so on. It also dwelled on the economic aspects, anticipated direct and indirect benefits, plan allocations and budgetary aspects, manpower requirements, management structure and institutional mechanisms. It was a comprehensive exercise of numerous one-to-one sessions and three long stays in Delhi. I had Unnikrishnan as my right hand while navigating through this long process that lasted one and a half months.

The seven-member national committee submitted its final report on 16 March 2009. It had the unanimous endorsement:

The Committee strongly supports the Indian human spaceflight programme and unanimously recommends that the Project Report submitted by the Department of Space should be processed for further necessary approvals so that the programme, which is a part of the approved 11th Plan, could be implemented without delay.

This was a red-letter day for ISRO—albeit a short-lived one. Ironically, the fiftieth Lok Sabha general election was notified a fortnight before the submission of this report. Wisdom and propriety prevailed as it was decided to take up the project for approval after the new government was in place.

The Way Forward

Contrary to the apprehensions of Delhi—that the present euphoria of the manned space mission might take away ISRO's focus on delivering operational space systems and services—we kept surging forward. After the launch of Chandrayaan-1 on PSLV-C11 in October 2008, the next target for VSSC was to fly the Geosynchronous Satellite Launch Vehicle (GSLV) with the Indian cryogenic stage.

GSLV had been spelling trouble for us since its very first flight. Although originally derived from the workhorse vehicle, PSLV, GSLV always had its own set of complexities. PSLV and GSLV did share a core 139-tonne solid motor. This core motor was complemented by a set of strap-on motors to form the first stage of both the rockets. While PSLV had six strap-on motors with a nine-tonne solid propellant each, the GSLV had four liquid strap-on motors with forty tonnes of propellant each. Both launchers had identical liquid second stages with minor modifications. But the main distinction came in the last stages. While PSLV had two more stages (solid and liquid sequentially), GSLV's last stage was a cryogenic stage that used liquid hydrogen and liquid oxygen as propellants. It was the high efficiency of this final stage that set the two launchers apart and gave GSLV almost double the payload capacity of PSLV. ISRO had started developing its own cryogenic stage in 1992, and also procured seven cryogenic stages from a company in Russia, for immediate operational use.

GSLV huffed and puffed through its first flight in April 2001. The next two missions were spot on and hence a lot of hope was riding on the GSLV-F02, the third flight, to take-off on

10 July 2006, carrying the INSAT-4C satellite. Initially, the take-off seemed flawless, but one of the four strap-on stages was too vigorous at the start and then its engine soon ceased to operate. The outcome was ghastly. The GSLV-F02 broke apart and the fragments fell on the coast near Sriharikota; an appendage of the launch pad was damaged too. The emergency response teams sprang into action and contained the damage and casualties. Mr Madhavan Nair was up front in announcing the failure on national television. The malfunction of a fluid control component was pinpointed as the root cause and corrective measures were taken. Within the next fourteen months, on 2 September 2007, GSLV-F04 came back with a successful flight, launching the INSAT-4CR communication satellite. Everything looked fine initially, but even this time the control system of one of the strap-on stages failed, which led to the GSLV's erratic movements. It resulted in the misjudgement of its target orbit for injection of the INSAT-4CR satellite.

As director of NRSA, I had observed these two tricky flights and their aftermath at close quarters. The failure modes of the gas motor that got stuck during the early part of the flight were still to be fully understood and confirmed when I took the helm of VSSC. Naturally, GSLV was one of the prime areas of concern for us.

Our colleagues at LPSC were doing their best to realize an Indian cryogenic upper stage built around our own cryogenic engine, as a replacement for the Russian stage used for the last five flights. The engine had been qualified through a specified duration of ground tests, in the year 2003 itself. The close equivalent of the flight cryogenic stage (that included the engine, fuel tanks and fuel circuitry) was also qualified through a sequence of prescribed ground tests by November 2007. There were still a few minor technical issues to be resolved for the maiden flight of the Indian cryogenic upper stage. The crucial acceptance test of our flight engine went off well on 18 December 2008. The project director of GSLV, G. Ravindranath, and the project director of the cryo stage, Mohammad Muslim from the LPSC team, went full steam ahead for the GSLV-D3 flight. On the bright and sunny afternoon of 29 September 2009, the chairman, Mr Madhavan Nair, flagged off the cryogenic stage from Mahendragiri for Sriharikota. This was the culmination of ISRO's relentless hard work of seventeen years going to the ultimate test bed.

Besides the GSLV-D3, the flight of two PSLVs with a month's gap was mandatory in 2009. The PSLV-C12 for launching a microwave remote sensing satellite RISAT-2 and PSLV-C14 for launching Oceansat-2 had to be prepared simultaneously; besides the induction of a new generation of avionics, both for mission computing and launcher telemetry. Many of us at VSSC and LPSC could not rationalize this sudden thrust for the twin missions of PSLV at such short notice as the GSLV-D3 deserved ISRO's full might. I vividly remember the discussions of January 2009 at the headquarters and the consultative process in VSSC for a seminal decision to appoint P. Kunhikrishnan, an avionics expert from Systems Reliability Entity, as an associate director of the PSLV programme. Kunhi, as we fondly called him, later turned out to be the most successful project director of PSLV, with more than a dozen successful flights under his belt.

We had a successful launch of PSLV-C12 on 20 April 2009. However, premature closure of the Chandrayaan-1 mission due to the failure of a device prepared us for a rigorous audit of all avionics packages before clearing the PSLV-C14 that was to launch the Oceansat-2 satellite. The successful launch on 23 September 2009 was special for me. The payload requirements for Oceansat-2 were driven by the user demand, and as the director of INCOIS, I had then teamed up with Shailesh Nayak and V. Jayaraman to specify these requirements. All three of us have changed our domains since then, but we shared key roles in this mission. Shailesh was the director of INCOIS; Jayaraman was the director of NRSA and I was there at the Mission Control Centre with PSLV.

While these operational programmes were going at full speed, ISRO's next generation launch vehicle, LVM3, was being developed. The programme was approved by the government in 2002. The capacity of LVM3 was almost twice that of GSLV, and hence it had the prospect of reducing 'cost per kilogram mass to orbit' by a great margin. The original commitment for its first flight was in 2007–08, but we still had a long way to go. Ramakrishnan, its first project director, passed on the mantle to another veteran, N. Narayana Moorthy, and took charge as chairman of the project management council. I contributed as a member. LVM3 was distinctly different from the PSLV and GSLV. LVM3 had, in its core, a liquid stage made of a pair of liquid engines that used 110 tonnes of earth-storable liquid propellant. Basically, this was an upgrade from the liquid engines used in the second stage of PSLV and GSLV that used 40 tonnes of propellant. LVM3 also had two solid strap-on motors, tied with the core stage, each of which had 200 tonnes of solid propellant. These strap-on motors, much heavier than the core motor of PSLV or GSLV, had to perform like identical twins in LVM3, otherwise it could swing. The upper stage of LVM3 had a high thrust cryogenic engine, producing almost double thrust level and double the amount of propellant loading. But of course, the cryogenic stages used in GSLV and LVM3 had different working principles and other engineering aspects.

Obviously there were a number of tasks to be completed across centres of ISRO. In tandem with the Liquid Propulsion Systems Centre and the Satish Dhawan Space Centre, we gave momentum to the development of the critical subsystems of this new launch vehicle.

At the end of autumn in 2008, we still had our fingers crossed for the success of Chandrayaan-1 entering the lunar orbit; the nation was celebrating Diwali, the festival of lights. The spacecraft teams at Bengaluru and Byalalu were deeply immersed in monitoring the health of the Chandrayaan-1 spacecraft and using the novel strategy to reach the moon by raising the orbit of the spacecraft around earth. Friends from the media at Thiruvananthapuram had been calling repeatedly to check up on

Chandrayaan-1. After a busy day at VSSC, I was just packing up when the phone rang. It was H.N. Madhusudhana from the headquarters in Bengaluru. 'Sir, hearty congratulations. We are all so happy,' he said excitedly. For a moment I thought he was referring to the day's operations at Byalalu. But Madhu continued, 'I just received the order appointing you as member of the Space Commission, effective from 24 October 2008. I am faxing the order to you immediately.' I was thrilled to bits. I immediately called up Mini and told her the news and then shared it with my close aides in the office, who were equally excited.

I had been familiar with the workings of the Space Commission since the 1980s, as I had assisted the former chairmen—Prof. Dhawan, Prof. Rao and Dr Rangan. On many occasions, I had appeared before the commission to make presentations on different subjects. In the recent past, the chairman, Mr Madhavan Nair, had invited me to make a presentation before the commission, in April 2008, on the Human Spaceflight Project Report.

My elevation as member of the commission was a great recognition, an honour. It was a sign that the management's faith in me had been reiterated. I knew if I worked hard, I would be considered for bigger responsibilities. I visited the Space Commission, as a member, on 24 December 2008, armed with my perspectives in relevant areas and articulated them proactively.

I must mention the name of one of my greatest well-wishers here. Dr T.K. Alex, a senior colleague at ISRO from the early 1970s, had taken over as director, ISRO Satellite Centre, after successfully establishing and running ISRO's Laboratory for Electro-Optics Systems (LEOS) for two decades. We had guessed by then that we were both being considered for the post as we were invited to the commission, in August 2008, to discuss 'India's Space Vision 2025'. The next morning after I received the order, I got an SMS from Dr T.K. Alex that read, 'You are one step closer to the destination.' Soon after this I received the good news that I had been selected for a fellowship at the Indian National Academy of Engineering. Later, I came to know that Dr Alex had made a ten-minute presentation at the INAE council meeting for selection of new fellows on my candidature. Dr Alex has remained a great source of strength and support. We grew as good friends and remain so even now.

My career too had progressed smoothly since my re-entry into the ISRO family in November 2005 in the grade of 'outstanding scientist'. The past setbacks were forgotten, as I was elevated to the grade of 'distinguished scientist' on 29 February 2008. The best was yet to happen. On 16 April 2009, I was at Sriharikota for a prelaunch review of the PSLV-C12, when I received a government's order, promoting me as 'distinguished scientist' in the apex grade, equivalent to the grade of Secretary to the Government of India. I promised myself to perform and stand out to demonstrate my potential and skills required for the coveted position at the high table of the Space Commission, the Department of Space and ISRO.

Another opportunity came on 12 August 2009, when the Astronautical Society of India held a one-day workshop at the India Habitat Centre, Delhi, on '21st-Century Challenges in Space'. The minister of state, Mr Prithviraj Chavan, was the chief guest, Mr Madhavan Nair presided and Dr P.S. Goel gave the keynote address. In the first technical session, I was assigned to talk on 'Space Transportation Systems', Ramakrishnan gave a presentation on 'human spaceflight', Dr T.K. Alex on 'spacecraft systems' and Dr Navalgund on 'Space Applications'. I had prepared well and did my best to complete my presentation in the allotted twenty minutes to the satisfaction of Mr Prithviraj Chavan, Prof. Rao and Dr Rangan. To many participants, that workshop looked like a 'town hall interview' for the top slot of ISRO.

After two weeks, I had the honour to celebrate my sixtieth birthday (albeit four days earlier) in the august presence of the Prime Minister Dr Manmohan Singh. It was not quite a celebration as I was invited to his residence along with the chairman, Mr Madhavan Nair, for the inauguration of the Indian Institute of Space Science and Technology (IIST). On that occasion, the PM also presented ISRO's Performance Excellence awards to twenty-five high performers in all age brackets before a distinguished audience. I introduced each awardee succinctly. In that short yet very important function, the PM was flanked by the chairman and me throughout. My well-wisher, benefactor and a former member (finance) in the Space Commission, Sisir K. Das, rightly observed, 'Radha got the golden opportunity and he used it well.'

Mr Madhavan Nair was completing his extended tenure as he turned sixty-six on 31 October 2009. There were enough reasons to believe that I was in a favourable position in case a change in the leadership at Antariksh Bhavan was initiated by the government.

The ball was in my court. I felt that I had to be ready for a round of interaction with Delhi at any point of time. I started to put down my thoughts for a road map for ISRO. I spoke to some of my colleagues to understand some nuances. But I had to be cautious, this preparation had to be discreet and I did not want to be labelled as unduly ambitious. I gathered a team of three of my younger colleagues-cum-conscience keepers who, in my judgement, had the mettle to think on that level. We prepared a vision document as well as a presentation over a few weekends in my apartment at Kowdiar in Thiruvananthapuram. This document was later titled 'ISRO Saga Forward' and I established it as my agenda in my new role as the chief of the Indian space programme.

19

SEVENTH CHIEF OF ISRO

In the wee hours of 22 September 2009, the day before PSLV-C16 was to launch ISRO's Oceansat-2 satellite, the chairman, Mr Madhavan Nair, and I visited the Balaji shrine at Tirumala Hills. This was a custom that we had been following before all the launches in my tenure as director, VSSC. After the darshan, we halted, for a while, at the guest house of the National Atmospheric Research Laboratory at Tirupati. I had with me a freshly printed copy of the 'Road Map to ISRO'; I had prepared this as Mr Madhavan Nair was retiring the next month and the search for the seventh chief had already begun. A search-cum-selection committee had also been formed for this purpose. Being the most senior director (in terms of grade) and a member of the Space Commission for almost a year, I felt I needed to prepare for the next possible elevation. Of course my guiding principle, 'no preparation goes to waste' kept me going. However, I intended to present the first copy to the chairman. I wanted to seek his blessings, as he was like my elder brother. He went through the document carefully and wished me luck. He also suggested that I should make a brief presentation on ISRO to the vice president of India, who was to arrive at Sriharikota to witness the launch. We finished our breakfast and proceeded to Sriharikota which was just a couple of hours drive from there.

As per schedule, the vice president, Mohd. Hamid Ansari, landed in Sriharikota in the afternoon of 22 September 2016. A briefing session on ISRO and the significance of PSLV-C16 and Oceansat-2 missions was held at the Kalpana guest house the same evening. The chairman introduced his ISRO council and senior functionaries to the vice president and left the floor for me to make the presentation.

The next day, 23 September 2009, soon after the launch, the vice president delivered a stirring speech that enthralled the nation. This was required for the staff of ISRO as they had been on the receiving end after the premature termination of the Chandrayaan-1 mission. The next day, NASA made an announcement about the discovery of water molecules in the lunar atmosphere. That created national euphoria. We returned from Sriharikota on a happy note.

In the first week of October, a discreet telephone call from the joint secretary of DOS, seeking personal information about my past service, hinted that I was in the reckoning for the ensuing succession.

Business continued as usual at Antariksh Bhavan and VSSC. In the third week of October, the yearly ISRO council meeting was held at Antariksh Bhavan with all senior functionaries deliberating and deciding on the programme and budget for the next eighteen months. The chairman made a media announcement on setting up a National Institute for Climate and Environment Studies at Bengaluru under the Department of Space.

Soon after, Mini and I decided to take a short break, and reached our home town on 24 October. I vividly remember that day, a Saturday. We started the day with a visit to the Koodalmanikyam temple located close to our ancestral houses. Bharatha, the deity there, is described in the Ramayana as the epitome of devoutness, righteousness, generosity and sacrifice. Next I called on Ambujam Ma'am, my beloved teacher from primary school. She was excited to introduce me to her relatives. After that we had lunch with my aunt at my ancestral house, took her blessings and immediately left for the Sree Krishna temple in Guruvayur. On the way, we stopped at Mini's uncle's place at Mundur, a remote village located between Thrissur and Guruvayur. Suddenly my mobile rang. It was A. Vijay Anand, the joint secretary of DOS, who exclaimed with a lot of cheer, 'Sir, hearty congratulations! We . . . order. You are going to . . . shall I fax . . .'

The intermittent mobile network kept me guessing. I replied, 'Many thanks. Can you read it out loud, it is difficult to hear you or locate a fax machine where I am now.' Vijay Anand too perhaps could not get me completely, but he did not relent. This time he managed to keep his statement as short as possible. He said, 'Sir, you must see it, give me a fax number.' I hesitantly agreed to message an accessible fax number.

My friend, K.B. Suresh, the deputy superintendent of police, was to join us at Guruvayur. We quickly rang him up and asked for a fax number that could be accessed by either of us. His first suggestion was to send the fax to his police station. But within five minutes, he messaged that the fax machine was out of order. He suggested an alternative number. I forwarded it to Vijay Anand.

Within the next thirty-five minutes, we reached the temple guest house at Guruvayur. I took a bath and got ready to proceed for darshan. Suresh had also arrived by then. I heard the manager of the temple office telling Suresh, 'A fax message has just arrived from Bengaluru. After the temple visit, I will bring it to Radhakrishnan Sir's room.' Suresh proclaimed in his bass voice, 'It must be important, we want to see it right away.' And then the piece of paper arrived and took away all the confusion. It put me in a place adorned only by a handful of brilliant scientists of India. The order, issued by P.K. Misra, secretary to the appointments committee to the cabinet, stated:

The Appointments Committee of Cabinet has approved the appointment of Dr. K. Radhakrishnan, Distinguished Scientist and presently Director, Vikram Sarabhai Space Centre, as Secretary, Department of Space and Chairman, Space

Commission for a tenure up to the age of 64 years vice Shri G. Madhavan Nair, on completion of his tenure on 31st October 2009.

Receiving that order in the holy precincts of the temple was a sheer coincidence; but for me it was a divine approval.

I placed the appointment order on the sacred steps of the Sree Krishna shrine (*sopana mandapam*). I bowed before him in reverence and gratefulness for making me worthy of this duty, and wholeheartedly prayed to be guided, blessed and protected by him while doing my duty. Indeed, I experienced his divine presence around me during my tryst with tribulations, trials and triumph over the next five years.

By then, the news had broken out in the media. Newspapers carried pictures of the next chief of ISRO at Guruvayur. An elated Guruvayur embraced me; Thrissur also expressed its pride and celebrated the success of an alumnus of its own engineering college. My home town, Irinjalakuda, rejoiced too as their boy next door had made it big. I came back to Irinjalakuda three months later for a civic reception where 5000 students made a human chain to greet and felicitate me in the presence of Kalam Sir.

The last week of October 2009 turned out to be a really hectic one. Mr Madhavan Nair was wrapping up at Bengaluru and Delhi. On 29 October 2009 at Delhi, he chaired a few meetings of apex bodies including the Space Commission and Board of Management of IIST that had important business to transact and decisions to take. A warm farewell function was arranged for him on behalf of the Space Commission, in which I was welcomed to my new role. V.V. Bhat who was soon to become a member (finance) of the Space Commission joined us for dinner after the event.

Thirty-eight years and six months after I stepped into the building of VSSC as a new recruit, I took over as the chief of the Indian space programme on 31 October 2009. Mini, my brother Sivadasan, and sister-in-law, Usha, attended the handing-over ceremony with a sense of pride and delight. Subsequently,

P.S. Veeraraghavan took over as director of VSSC. Both of us escorted Mr Madhavan Nair to a room reserved for him at VSSC, next to the director's office. The transition had been seamless.

That forenoon, VSSC organized a simple function to lay the foundation stone of a space observatory on a piece of land allotted by the Kerala government at Ponmudi, a small hill station in the Thiruvananthapuram district. The gracious presence of Mr T.K.A. Nair, principal secretary to the prime minister and member of the Space Commission, was heartening. His words of wisdom to me were invaluable. He had known me since 1994, when I was engrossed in remote sensing applications and he was a prime promoter of it from Delhi.

By afternoon, Mini and I left for Bengaluru, accompanied by V.S. Hegde, the director of the Earth Observation System Programme at the ISRO headquarters. He was sent as an emissary of the headquarters to receive the new chairman. Somanath joined us at my request. Mrs Ramadevi also joined us to witness me taking over the 'chair'; the teacher in her must have been proud. Srinivasa Kumar (of INCOIS) promised to come down from Hyderabad to greet the new chairman.

The one-hour flight from Thiruvananthapuram made me introspect. I was about to step into the shoes of an array of illustrious space scientists in the country, who had made a mark for themselves in the space fraternity of the world. I was awed and felt reverence towards my predecessors. Of course, a kind of self-doubt did seep in. But I knew that with my techno-managerial skills, I was well equipped to appreciate the space programme as a complex, high-risk, multidisciplinary system. I felt confident in my proven track record of performing in different domains under difficult circumstances. I felt good for having a career in an assortment of domains. I had garnered maximum experience in space applications, and of course my knowledge on space transportation systems would be a big advantage. With these, I was conversant in two of the three most important domains in ISRO, i.e. the space transportation systems, spacecraft systems and space applications. Also, I had the invaluable exposure of closely working with three chairmen from 1981–97. During this time, I gathered rich experience and insights into the relevant scientific, technological, programmatic, financial, managerial and geopolitical issues. I had worked closely with Delhi during my tenure at the headquarters and at INCOIS/NRSC. I felt confident about bringing new value propositions to the table. I was excited with the prospect of my new role to extract excellence from the 16,000-strong human capital of ISRO. I was confident and raring to go.

On the evening of 31 October 2009, I stepped into Antariksh Bhavan as the seventh chairman of ISRO. The ISRO headquarters at Bengaluru had been my second home for sixteen years, since 1981. An avionics engineer turned manager had bloomed here. Mini was beside me as a silent observer, recounting the turn of events after which I had left the campus twelve years earlier. The red-carpet welcome from the main gate was truly touching. My colleagues were buoyant at my re-entry into the headquarters. I carried with me a copy of my appointment order, my 'Road Map to ISRO' and a framed photograph of Prof. Satish Dhawan that I had kept with me for many years. I was proud to be his last doctoral student and got a tremendous amount of strength by remembering his philosophies. The photograph kept reminding me, 'Never do anything that I won't do.' These were his words when I was about to leave for my PhD at IIT Kharagpur. I paid my deepest respect to his soul and occupied the precious seat for the very first time.

The scientific secretary, A. Bhaskaranarayana, put a paper on my table for my signature as I had asked him to; it pertained to the due process for inviting Mr Madhavan Nair as the Vikram Sarabhai distinguished professor for a term of four years, coterminous with my tenure. Later, I exchanged pleasantries and old memories with senior colleagues from DOS and the headquarters, over a cup of tea. I called up Dr Rangan (the then member, science, Planning Commission) to seek his blessings. He told me he was always there to guide and support me. Afterwards, along with Mini, I visited Prof. U.R. Rao and Mr Madhavan Nair to receive their blessings.

20

SETTING THE STAGE

At 8 a.m. sharp, on 1 November 2009, a gentle knock on the door of my suite at the ISAC guest house caught my attention. I was getting ready for my first day at Antariksh Bhavan as the new chairman of ISRO. From the other side of the door, sub-inspector (CISF) Nitin Kumar Sinha greeted me. 'Good morning sir, we are ready to start.' I was delighted to see Nitin, my personal security officer for nine months at VSSC, taking care of my security at Bengaluru as well.

Dhanya had gifted me a Louis Philippe blue formal shirt and a pair of navy blue formal trousers to wear on this special day. The guest house at ISAC became my home for the first few months, because the official residence, Vyoma, in RT Nagar, Bengaluru, was being readied after a gap of almost six years. Mini preferred to stay with Dhanya and her mother to tie up some personal issues, and that allowed me to fully focus on office matters.

On the very first day, after reaching the office, I was elated to see that Raghunatha Rao, a trusted personal assistant of a decade at Antariksh Bhavan, had been posted to work with me besides the veteran of the chairman's office, Mrs Yashoda and her two aides—Sreedhar and Padmanabha.

Upon arrival I was informed that Prof. U.R. Rao had asked for an appointment in the forenoon. I felt a bit embarrassed, immediately checked about his availability and rushed to his room to meet him. Prof. Rao was heading a peer review committee, formed by the prime minister's office, to review the recently terminated Chandrayaan-1 mission. Prof. Roddam Narasimha and Dr V.S. Ramamurthy were the other members. In the hour-long meeting, Prof. Rao shared the key observations from this review and alerted me on a set of curative steps to be followed in the future.

The main constituents of the ISRO headquarters were the programme offices dedicated towards launch vehicles, satellite communications, earth observations, space sciences, budget and economic analyses, etc. Each of these were headed by a programme director. The scientific secretary would officiate as the head of office of the headquarters. At that time, A. Bhaskaranarayana, a Vikram Sarabhai distinguished professor (after his superannuation), had been functioning as the scientific secretary and director of the satellite communications programme. V.S. Hegde was heading the earth observation programme. H.N. Madhusudhana, whom I had handed over the charge of director of budget and economic analysis in 1997, was continuing in the same role. D.R. Suma headed the launch vehicle programme. Antariksh Bhavan also hosted the secretariat of the Department of Space that had a few senior bureaucrats to head key functions. Additional secretary, G. Balachandran, an IAS officer of the West Bengal cadre, and joint secretary of the Indian Revenue Service, A. Vijay Anand, were also part of my core team at Antariksh Bhavan.

I discussed 'The Road Map to ISRO' with my core team; their suggestions and observations were taken on board. The scientific secretary, Bhaskaranarayana, suggested that we rename the document 'ISRO Saga Forward'. It was appropriate and I happily accepted it. I expressed my wish to present the road map to the entire ISRO family in the farewell function for Mr Madhavan Nair, planned the next day. Incidentally, Prof. Rao and Dr Rangan also made themselves available for the occasion. That was to be a special one.

In the afternoon, Dr Alex, director of the ISRO Satellite Centre, made a surprise visit with a bouquet and insisted that I visit his centre that day itself. He gave me an informal briefing on the key issues on the satellite front and took me to the magnificent satellite integration and test complex, famously known as 'ISITE'. During the coming years, I heavily relied on the technical and managerial judgement of Dr Alex, a very close friend and well-wisher.

Apart from him, my ISRO council was studded with stalwarts who headed the other six major centres and units of ISRO. The directors—Dr Ranganath R. Navalgund at the Space Applications Centre, P.S. Veeraraghavan at VSSC, M.K.G. Nair at LPSC, M. Chandradathan at SDSC-SHAR and V. Jayaraman at NRSC—formed the core of my management team. Except Dathan and Jayaraman, the rest had held senior positions to me for long stints in their respective domains. They have all significantly contributed in my stint as chairman, ISRO.

Dr B.N. Suresh, with whom I had a long professional association, was officiating as the founding director of the Indian Institute of Space Science and Technology. Antrix, a government-owned entity under DOS, had K.R. Sridhara Murthi as its managing director.

The tradition of seamless transition at the helm of ISRO continued this time as well. I had been in the Space Commission for a prenatal period of one year. My predecessor continued with us as a Vikram Sarabhai distinguished professor, operating from a room opposite mine at Antariksh Bhavan. As a symbol of respect, I offered him the chairman's car and I opted for a smaller one for myself. Mr S.K. Das, a former member (finance), was gladly available as an honorary adviser; he had been my well-wisher since his days as the joint secretary at DOS in the early 1990s.

The next afternoon, Antariksh Bhavan saw an emotional farewell and rousing welcome at the same function. All centres and units of ISRO and DOS were hooked up through satellite link. Prof. U.R. Rao and Dr Rangan flanked Mr Madhavan Nair and me on the dais. As discussed, I made a presentation on the 'ISRO Saga Forward'. I felt that my predecessors were happy and convinced about the future.

A Leader's Responsibility

Having communicated my ideas and plans, it was my duty to get consent from Delhi. Since inception, ISRO, the Department of Space and the Space Commission were under the direct charge of the prime minister. On 9 November 2009, I briefed the minister of state (PMO), Mr Prithviraj Chavan, the national security adviser, M.K. Narayanan, the principal secretary to the prime minister, T.K.A. Nair, and the cabinet secretary, K.M. Chandrasekhar about the current programme and priorities. All of these gentlemen were ex officio members of the Space Commission. On my invitation, Mr Prithviraj Chavan agreed to fly down to Thiruvananthapuram and interact with engineers at VSSC and LPSC (he visited us from 2–3 December 2009). Mr T.K.A. Nair fixed up my first meeting with the prime minister the very next day.

My first briefing on ISRO to Prime Minister Dr Manmohan Singh took place at 1230 hours on 10 November 2009 at the PM's official residence, famously referred to as '7RCR'; the meeting lasted for thirty minutes. When the door opened for me to enter his cabin, I did not find the PM in his chair. To my pleasant surprise, he was just next to the door, waiting to greet me. Such magnanimity and grace remains inscribed in my memory. The national security adviser, M.K. Narayanan, joined us for the briefing. I made a presentation on 'ISRO Saga Forward'. The PM listened carefully and asked a couple of questions in between. I got the green signal from the highest office for my plan to take ISRO forward during the next four years. The PM showed a lot of interest in the preparation for the impending GSLV-D3 mission with the Indian cryogenic stage. I asserted my proposal for a comprehensive technical assessment by a national panel of eminent experts and highlighted the imperative for their unequivocal technical endorsement, as a prerequisite for authorizing such a high-stake national mission. We discussed the profile of the possible experts. The prime minister, concerned about the recent premature termination of the Chandrayaan-1 mission, appreciated this initiative and expressed his firmness on transparency and accountability. The discussions concluded with a reminiscence of professionalism and transparency during Prof. Satish Dhawan's regime as well as some advice to me to groom future leaders. It was heartening to note that the PM advocated the space programme, and he himself had contributed to the Indian space programme as a member (finance) of Prof. Dhawan's Space Commission.

The PM's advice of identifying and grooming the future leaders struck a chord. I acted quickly on his advice and constituted an 'ISRO Strategy Group' of 225 young high performers through a collective process along with the centre directors. The intent was to encourage these youngsters to widen their horizons, to explore newer areas of space technology and applications, work towards cost-effective solutions, and build scenarios and processes by which a future vision for the Indian space programme could be evolved. I invited my mentor, Mr Y.S. Rajan, to steer the process of mentoring the ISRO Strategy Group. That was the beginning of an institutionalized process to develop future leaders at different levels.

But programmatic development was my top priority. When I took over as the chairman, the imminent priority was to test the Indian cryogenic stage on board GSLV. A successful flight would have made ISRO self-reliant in launching its own communication satellites with a mass of up to 2200 kilograms. After the mandatory ground tests, one cryogenic engine, earmarked for the flight, was cleared in December 2008 for further steps of assembly. And soon the cryogenic stage (comprising the engine, propellant tanks and fluid circuitry) was assembled and made ready at the Mahendragiri facility. We were eagerly looking forward to the GSLV-D3 flight in which this cryogenic stage was to be tested. At that point, in 2009, we could have prepared ourselves for the all-important GSLV-D3 flights, but our efforts were hindered by an urgent order from the headquarters that mandated us to demonstrate the capability and immediately fly two PSLV flights within a span of one month. We still went ahead with the launch campaign of GSLV-D3; the lower stages were assembled in May 2009. The cryogenic stage—already realized and ready at Mahendragiri—was sent off to Sriharikota in September 2009. By then I was moving up as the chairman and wanted to take up this challenge at the start of my regime itself.

My well-wishers told me to first take up the relatively easier and safer PSLV mission rather than the GSLV, that too with the additional risk of the maiden flight of the Indian cryogenic stage. By then, the PSLV had had a golden run of fifteen consecutive successes, and I had seen five of them through as director of VSSC. I appreciated their concern for me. But I strongly felt that just for the sake of having a successful first mission as chairman I should not delay the flight test of GSLV-D3, more so because the Indian cryogenic stage was truly a national imperative.

There were residual technical issues in the GSLV-D3 that had come up after the Flight Readiness Review (FRR) and the Mission Readiness Review (MRR), the customary prelaunch review forums. A couple of these issues were crucial. For example, whether the liquid hydrogen turbopump, a key element of the cryogenic engine, would rotate at minus 253 degrees Celsius or not; whether ignition of the main cryogenic engine and its two steering engines would be smooth in the vacuum conditions at an altitude of 130 kilometres. These had not been addressed so far, as we did not have a test set up to simulate these conditions, though analytical inputs were available. A striking revelation on the frailty of the thrust frame of the cryogenic stage surfaced on 20 November 2009.

The idea of a comprehensive technical review of GSLV-D3 by the National Panel of Eminent Experts was vindicated. The national panel was set up with three of my predecessors and several former centre directors of ISRO. These experts were either involved in cryogenic technology development since the beginning or were still active members of multi-tier preflight-review forums. The panel also included Prof. Roddam Narasimha and nearly twenty eminent aerospace specialists and academicians of the country. The current group of project functionaries and launch vehicle specialists of ISRO were also members by default.

The first session of this comprehensive technical assessment was held on 20 December 2009 at Antariksh Bhavan. The project teams captured the essence of the discussions and listed the action points therein. Technical teams were set up to address a dozen issues through analysis and tests on the thrust frame. An assessment of the life of the stages and subsystems of GSLV-D3 (stacked since May 2009) was tasked through a series of revalidation and verification tests. While these teams pursued the actions in all earnestness, I wrote to the national panel on 28 December 2009 to elicit any further concerns they might have as well as to request analytical commentaries on the issues that had surfaced. I did not want to leave any stone unturned; our green signal for the flight had to be based on the best understanding and judgement available nationally.

Meanwhile, on 30 December 2009, Prof. U.R. Rao, Prof. Roddam Narasimha and Dr V.S. Ramamurthy presented their report on the performance of the Chandrayaan-1 mission to Prime Minister Dr Manmohan Singh. As instructed by the PMO, I was present at the meeting. ISRO was advised to take a set of corrective measures for future space missions, primarily in the areas of thermal design and quality control. The final advice handed over to me was:

> The Committee compliments ISRO and the Department of Space for its long record of being scientifically open and transparent, and would like to strongly urge to continue with and even enhance this policy in the future.

At the close of this session, Prof. Rao and Prof. Narasimha joined me to update the prime minister about the outcome of the twentieth review of the GSLV-D3 held in December; they too reiterated the imperative of such a process in this mission.

The national panel reassembled on 28 January 2010 and synthesized results of the tests and analysis as well as specific responses from its members. At its close, the national panel said 'no major concerns' and gave a green signal to go ahead. Seven out of the eight flagged issues had been addressed and resolved. On the fuel booster turbopump, an expert team was tasked to reconfirm failure modes at cryogenic temperature, if any. With this, I handed over the baton of the GSLV-D3 mission for the normal prelaunch review process of FRR, MRR and the Launch Authorization Board.

Amidst all these efforts to ensure a successful flight test of the Indian cryogenic stage on board GSLV-D3, I was amazed to receive a strange message on 11 January 2010 from my predecessor, starting with, 'Now seeing the progress of events, I would like to express my concern on the time delays on account of the managerial process.' The message concluded by saying, 'You may like to revisit the decision-making process which will help to progress a critical mission like GSLV-D3 and ensuring its success.'

I took it in my stride as advice from a well-wisher and benefactor. However, I had full conviction that such a consultative process was inevitable before giving a green signal for this much-awaited flight, a responsibility that I inherited.

Meanwhile, on 24 January 2010, ISRO crossed a major milestone. The S200 solid propellant rocket stage developed by VSSC and SDSC for the advanced launcher, LVM3, was successfully tested on the ground. That was the first test of the third largest solid rocket of the world (only next to the solid boosters of the Space Shuttle of the US and the P230 solid booster of the Ariane-5 of Europe). Prime Minister Dr Manmohan Singh told me on the phone, 'You have made an auspicious start. This is yet another feather in ISRO's cap.' I thanked him and said, 'Dathan, director of the Satish Dhawan Space Centre and the first project director of this rocket stage, is with me.' I then requested Dathan to brief the PM and he did.

We logged two more achievements in the first week of March 2010: ISRO's advanced sounding rocket, developed at VSSC as a test bed for the air-breathing propulsion technology, went through a successful flight test; the L110 liquid rocket stage, developed by LPSC for the LVM3 launcher, went through 150 seconds of tests on the ground. Within a couple of days at Delhi, the prime minister congratulated me as I sat with him to sign a memorandum of understanding (MoU) between ISRO and the Federal Space Agency of Russia on 'Joint Enterprises for Production of Satellite Navigation Equipment and Services for Civilian Users'.

All Roses Have Thorns

As in the case of any large organization, transition means inheritance of a legacy, cumulative gains and assets from the past generations. Also all these come coupled with a noble vision, a vibrant institution and a wonderful team. The baton is passed on from one leader to the next to pilot the organization to higher planes and explore new avenues to make a mark. Nevertheless, it does come with a caution of not perpetuating any weaknesses that might have crept in.

After a transition, the organization strives to sync with the new leadership as both come to terms with each other. It is normal that a few unfortunate team members fail to reconcile with the change. One can associate such incidents with some of the inner circle members of the previous regime. They tend to become erratic. I have seen this happening at ISRO headquarters during the previous transitions. Unfortunately, as a new incumbent, I had to face a larger dose of such consequences. The first salvo was fired on me in the first fortnight itself. And then I continued to face frequent spurts of it; some of which went to abominable levels. I had to deal with the situation, and I could do this quickly because of my strong bond with the rest of my team. Nevertheless, it did leave behind some bitterness in the system that could have been avoided.

But there were many more surprises for me in the first few months and not all were pleasant. On my first working day as chairman, V. Jayaraman, my successor at NRSC, briefed me on a major procurement contract, in which a foreign vendor had jacked up his bill by 60 per cent after the letter of intent had been issued. I felt it prudent to seek the views of my predecessor formally but his advice, on file, to pursue it was not tenable, to say the least. I decided to recommend alternative options to the government.

The idea of amalgamation of the five RRSSCs (which I had the privilege of establishing as the project director) with the NRSC was hanging for a long time, though it was a managerially efficient option for both. With the moral responsibility of a parent, I persuaded Y.V.N. Krishnamurthy, director of RRSSC, to amalgamate his office with NRSC. Both teams stood to benefit from the synergy of talents, resources and avenues for growth. That was a welcome change.

The next delicate moment appeared in the second week after my taking charge when source information on certain alleged malpractices and procedural lapses in the Antrix-Devas Agreement (of January 2005) were brought to my attention. After an internal review with the functionaries concerned, I appointed a single-man committee of Dr B.N. Suresh for a comprehensive review on 8 December 2009. Three senior functionaries from DOS, ISRO headquarters and Antrix were identified to assist him.

I had a firm belief that seventeen years after its inception, it was time that Antrix got a full-time chairman and managing director, and that the time was ripe for the ISRO chairman to divest the chairmanship of Antrix. This would have brought more operational flexibility and autonomy in the functioning of Antrix, which was mandated to market the space products and services of ISRO. However, I had to wait because of this new development, lest I was seen to be running away from the raging bull.

I soft-pedalled another ongoing initiative to fabricate crucial launcher hardware abroad and test the same in a foreign location. The proposition had its obvious techno-managerial vulnerability and associated geopolitical ramifications. I was surprised when my scientific secretary appeared to be unaware of this initiative.

The hurried pronouncement in October 2009 of plans to set up a National Institute for Climate and Environmental Studies (NICES) at Bengaluru had not been backed up with the prerequisite process of inter-ministerial consultations and clearances from the prime minister's panel on climate change. The inter-ministerial consultations, co-chaired by the environment minister and the member (science) of the Planning Commission in January 2010 came to the conclusion that NICES could be located temporarily at the National Atmospheric Research Laboratory at Tirupati, in view of the available facilities. Later, NICES was set up, permanently, at the National Remote Sensing Centre as a programme on national information for climate and environmental studies.

It is obvious that some of these decisions impinged on the ideas and thoughts of my predecessor. Time and again I had been blamed for overriding my predecessor's decision. But I always had the conviction that I took all decisions based on the merit of the cases and did my best to get a consensus of opinion from those who mattered. And I believe anyone in my chair would have and should have done so.

The Space Commission was reconstituted in January 2010. I had already been appointed as its chairman. Besides the minister of state (PMO), Mr Prithviraj Chavan, the principal secretary to the prime minister, national security adviser, cabinet secretary and finance secretary were members in ex officio capacity, whereas Prof. Roddam Narasimha, (a member of the commission since the late 1980s) continued in the same post. Dr R. Chidambaram (principal scientific adviser to GOI) and Dr T.K. Alex (from ISRO) were inducted. Additional secretary, Balachandran, became the secretary of the commission.

The next meeting of the Space Commission was held on 22 March 2010. Except for the MOS (PMO), who was out of station, all members participated in the fairly long meeting. I received accolades from the distinguished members of the Space Commission. I felt that I had passed the first test and met the expectations of my selectors.

Incidentally, 29 March was a proud day as I stood behind the podium at the Indian Institute of Management, Bengaluru, my alma mater, to deliver its thirty-fifth convocation address.

21

TIMES OF TRIALS

The festive charm of Vishu was subdued by the effervescence for the most-awaited flight test of India's cryogenic upper stage on the Sriharikota Island. On 15 April 2010, after eighteen years of concerted effort, India was finally on its way to join the elite club of countries such as the US, Russia, France, China and Japan, who owned the complex cryogenic technology.

Cryogenic stages have always been in demand, more so for the upper stage of a rocket that carries the satellite to higher orbits such as geostationary orbits. Compared to the solid and earth-storable liquid stages, used for the lower portion of a rocket, a cryogenic stage generates 45–60 per cent more force for every kilogram of propellant it burns. Thus it plays the central role of providing nearly half of the total velocity required for the rocket to deliver a satellite into the elliptical geostationary transfer orbit.

But a cryogenic stage is powered by the ignition of hydrogen and oxygen preserved in a liquefied state at minus 253 degrees Celsius and minus 183 degrees Celsius respectively. Such low-temperature fluids are difficult to handle; only a few materials can withstand such low temperatures. The initial ignition of hydrogen and oxygen is tricky, so is its sustenance. Gases from combustion of this fuel come out as hot as 3225 degrees Celsius, and hence the combustion chamber should be cooled down to 600 degrees Celsius to avoid melting it. A set of turbopumps, running at around 40,000 revolutions per minute pump the fuel into the combustion chamber. The list of complexities in the launcher is long and thus cryogenics have always been tough nuts to crack.

We kept our teams ready with contingency plans for any conceivable emergency during its first flight test. The chequered past of the GSLV launcher was sufficient to keep everyone on their toes.

While the integration of GSLV nominally took several months, the launch countdown operations took about thirty hours. Most of the crucial operations of fuelling the cryogenic stage would happen during the last couple of hours of the launch countdown. In the last ten–fifteen minutes of the countdown, the mission computer would take over the control of operations through an automatic launch sequence programme and the final set of commands would be issued to lift-off the GSLV rocket. Of course such automated clearances came after ensuring that the conditions essential for the flight were in place. The ignition of the cryogenic stage happened after the first five minutes of the flight.

The stakes were very high in the satellite front too. The GSAT-4 satellite, on board GSLV-D3, was an advanced communication technology demonstrator. A set of transponders operating in Ka-band (first time ever for ISRO) had prospects of broadband data connectivity. A transponder to broadcast navigation signals from GAGAN (GPS Aided GEO Augmented Navigation) too was built in, besides trying out electric propulsion. GAGAN was a joint programme with the civil aviation ministry that aimed at improving GPS signals used by aircraft during take-offs, landing or cruise phase. In hindsight, it was not probably the best decision that we took—to fly an important satellite like GSAT-4 in the test flight of the Indian cryogenic stage, on board GSLV. However, we did our best to ensure a successful launch, and after a clearance by the National Panel, we had no reason to not feel confident.

Embracing Eventuality

I had been camping at Sriharikota on and off during the previous fortnight as the prelaunch rehearsals and reviews were in progress. On 14 April, we visited the Tirumala Balaji temple for our customary early morning darshan before heading to Sriharikota for the twenty-nine-hour countdown. I was accompanied by a team from the headquarters. This was my first mission since I took over, and I was conscious of the fact that my colleagues and the public at large would judge my mettle at the hot seat in the Mission Control Centre. I did not have the luxury to burst or buckle.

I kept Prime Minister Dr Manmohan Singh in the loop about the progress of the countdown. His best wishes were important for the entire team and also for me. The very presence of a few doyens from the first generation of scientists from ISRO and my three predecessors—Prof. Rao, Dr Rangan and Mr Madhavan Nair—in the VVIP gallery at the Mission Control Centre was assuring.

Pant Sahib, who headed the Mission Readiness Review proceedings, was with us at the mission console. In the hot seat, I was flanked by the centre directors, the mission director and the other mission executives. The Brahm Prakash auditorium, located a few kilometres away from the main Mission Control Centre of SHAR, was overflowing with members of the media, who had gathered to witness history unfurl before them. And then there were thousands, if not millions, who were glued to their television sets to witness the same.

After a very smooth countdown of twenty-nine hours, GSLV-D3 majestically lifted off at 1627 hours (IST). The first round of applause came in, like a reflex action, from the viewing gallery. The four liquid strap-on stages (L40) and the solid core stage (S139) did their job perfectly and got detached after completing their actions. I heaved a sigh of relief—we had faced issues with these points even in 2006, when our flight had failed, and in 2007, when it had lost control. The second stage (GS-2) functioned well as usual. This stage, derived from the Vikas engine, was a confident candidate. The GSLV-D3 launcher followed its designated flight path and attained a velocity of 4.9 kilometres per second, almost half of what was required to inject the GSAT-4 satellite.

Nearly five minutes (293 seconds to be precise) after the lift-off, V. Narayanan (fondly called 'Mr Cryo'), one of the key persons responsible for developing the engine, confirmed over internal communication that the initial conditions required to start the cryogenic stage had been attained. Mission executives were glued to the mission consoles. The flight computer initiated an intricate sequence of nine discrete events over two seconds, essential to ignite the cryogenic engine. The announcement of ignition of the main cryogenic engine was received with long and thunderous applause.

Moments later I sensed that something was wrong. On my console I could track the time/altitude and time/relative velocity profiles of the flying rocket. It did not show any sign of altitude build-up and moments later the altitude curve started nosediving. Whispers turned into silence as we all knew that the rocket was losing its course. Ramakrishnan, director, projects, VSSC, came to me and whispered, 'It seems the Vernier engines are not functioning.' These Vernier engines, expected to control the attitude of the soaring rocket, became our first suspects. It was only a matter of time. GSLV-D3, sans the thrust, lost the battle against gravity and finally came crashing down along with its precious payload, the GSAT-4 satellite. From my console, I could see Ravindranath, the mission director of GSLV-D3, staring blankly at his monitor screen that was still showing the altitude/time profile of the rocket. The mission was over. I quickly jotted down my announcement and consulted Dr Alex and Veeraraghavan, the directors of ISAC and VSSC. I spoke from the mission console, 'The mission objectives of GSLV-D3 have not been fully met . . . Realization of the cryogenic engine and stage was a major accomplishment in itself, but we need to go a long way, and we would do that in the coming year.'

Failure is always an emotional moment. While some choose to go into a shell post the event, there are others who accept the fact and look for solutions. The space engineering folks, by default, need to be in the latter category. Failures are part and parcel of this trade and one needs to take it in one's stride. But for me, the situation was a tad harsher. I had just experienced the spectacular failure of my very first mission as the chief of the Indian space programme. My organization, the government and several millions had witnessed it. A trial by the media, who had already been hounding us for the premature termination of Chandrayaan-1, was inevitable. In that moment, I decided to take sole responsibility for the failure, present the facts to the government and the country, bear the brunt of the criticism, and more importantly hold the team together. I was composed to face the music. I offered a few words of reassurance to my colleagues at the mission consoles, before returning to the VVIP gallery to brief the dignitaries present.

No briefing was required. Prof. Rao put his right hand on my shoulder and his silent stare was eloquent. A teary-eyed Dr Rangan came up and hugged me. Mr Madhavan Nair looked completely disappointed as he consoled me. I left for my suite at the SHAR guest house to send a message to Delhi, before interacting with the raging media. An emotional Prof. Rao came to my suite, and silently sat by my side as I typed the brief report to the prime minister. Soon Prof. Rao left for Chennai to catch the late-night flight to Bengaluru.

The members of the media were restlessly waiting in the Brahm Prakash auditorium. Under normal circumstances, during such briefings, the chairman of ISRO would be flanked by centre directors and mission directors on the dais. This being a failure, I decided to face the media alone. My senior colleagues were requested to be seated in the front rows among the audience. It was my second media briefing as the chairman. At the close of the long interaction, a Chennai-based journalist asked me a personal question about how I felt about the maiden failure. In response, I just recited one of my favourite stanzas from the *Bhagavad Gita*, '*Karmanye eva adhikarah the; Ma Phaleshu Kadachana.*' It meant that you only have the right to work, but not dwell on the results. The fruits are not your motive. The media chose to stand firmly with us instead of being harshly critical, thereby enabling us to move forward.

After any launch, it had been a tradition at ISRO for the teams to assemble for a quick analysis of the flight performance, irrespective of success or failure. That day I dismissed the idea and chose to give a pep talk to boost the morale of the exhausted and grief-stricken engineers. I invited Dr Rangan and Mr Madhavan Nair to join us. Dr Rangan spoke eloquently with passion. Mr Madhavan Nair expressed his belief in the acumen of our engineers to solve the cryo riddle without any difficulty. At the close of the meeting, I told my colleagues to rest that night and start analysing the flight data with a fresh mind the next morning. I asked for a review at Thiruvananthapuram a couple of days later.

As I was returning, I saw a crestfallen Mr Cryo in tears. I pulled him along to my suite. I asked him a simple question, 'Narayanan, when your daughter started walking, did you never see her fall?' 'She got up and walked again,' he replied. I said, 'That is the lesson for us now. Yes, we gulped water while swimming, but we didn't drown.' He smiled. I added, 'Just lock yourself in the room and sleep. I am sure that you will have the answer after looking at the flight data with a fresh mind.'

On my way back to Bengaluru, I stopped at the Tirumala temple to offer prayers. Reactions in the print and electronic media suggested that the nation was positive and supportive. Several well-wishers rang up and wished us luck for our future missions.

The next afternoon, I got a call from Mr Cryo, who sounded positive this time. 'Sir, we know the cause. The "fuel booster turbopump" stopped about a minute after ignition. This failed our mission.'

On Sunday, 18 April, the team was ready at Thiruvananthapuram with the preliminary failure analysis. The hypotheses were: (a) the main cryogenic engine and two steering engines did start; (b) the fuel booster turbopump raced to its normal speed initially, then stopped abruptly within 1.5 seconds. Incidentally, a possibility of such a failure had been raised during the review by the National Panel and we had referred the turbopump for a review to an expert committee. We now needed to identify why it had stopped, demonstrate the hypotheses through ground tests and then take corrective measures.

A twenty-one-member team comprising experts from ISRO, national laboratories outside ISRO and academic institutions was set up under the chairmanship of S. Ramakrishnan (who was still director, projects, at VSSC) to carry out the analysis and submit its findings by the end of May 2010. A small group of young experts led by M.S. Suresh (also known as Mr Turbopump) was assembled at the ISAC guest house in Bengaluru; it applied itself to make a fishbone diagram of all possible failure scenarios. I requested the National Panel of Eminent Experts (that had earlier looked at the flight worthiness of the GSLV-D3 with this indigenous cryogenic stage) to review these recommendations and guide us on the future course of action. I flew to Delhi the same evening to make a detailed presentation to the PM and his senior colleagues at the PMO.

Within days, I went to our Liquid Propulsion Systems Centre at Mahendragiri and Valiamala (the LPSC Mahendragiri campus was renamed the ISRO Propulsion Complex in 2014) to interact with members of the teams who had developed the cryogenic engine and stage. My next destination was the Space Applications Centre (SAC) at Ahmedabad that had tirelessly worked on the Ka-band payload of the GSAT-4 satellite for five years but the satellite had not got a chance to enter the orbit. First, along with the director, Dr Navalgund, I met the teams that had built the

Ka-band transponders and conveyed the launch vehicle team's apologies to them. We had a session with the senior functionaries of the centre, and I finally addressed the entire SAC community for an hour or so. I recounted the failures that ISRO had faced from its inception and how, on each of these occasions, the teams had sprung into action to achieve success. I encouraged them to raise questions and they did; the interaction brought us closer. A similar process was followed with the teams who had built the GSAT-4 satellite at the ISRO Satellite Centre in Bengaluru. In the next few days, I interacted with mission controllers at Hassan who had been standing by to manoeuvre the GSAT-4 to the right orbit; they even had contingency plans in case the GSLV-D3 faltered within tolerable limits to place GSAT-4 into the specified orbit. I also made it a point to brief my colleagues at NRSC and ADRIN at Hyderabad.

A meeting of the Space Commission was held at Sriharikota on 2 May to boost the morale of the concerned teams. Besides the MOS (PMO), Prithviraj Chavan, Prof. Roddam Narasimha spent a day with the cryogenics and launch operations teams of GSLV-D3. We had the assurance that the government was firmly behind ISRO at this crucial juncture.

By July 2010, two plausible scenarios were identified for the failure of the fuel booster turbopump: (a) gripping at one of the seal locations and seizure of rotor; and (b) rupture of turbine casing due to excessive pressure and thermal stresses. A series of confirmatory ground tests had to be planned and a test facility had to be rigged up to test the turbopump in a low-temperature ambience. The possibility of contaminants in the fuel line could not be affirmed then.

Challenges Always Come in Groups

Even though I had declared our intent of flying the Indian cryogenic engine within a year, it soon became clear that the task would take more time, even with the team putting its best foot forward and the management giving it the top priority tag. A rough estimate of two–three years was made before declaring GSLV operational.

But then, we desperately needed an imminent launch of GSAT-5P by a GSLV, as the national position of satellite transponders was in a precarious state. The national demand for satellite transponders was growing for telecommunications (for governmental and non-governmental users), broadcasting (including direct-to-home services), education, developmental communication and disaster management support. Though our ten INSAT/GSAT satellites (including the INSAT-4CR launched in September 2007) provided nearly 200 transponders, we had leased some capacity from foreign satellites to bridge the gap. Moreover, about 25 per cent of our national capacity came from four of the existing INSAT/GSATs that were at the end of their operational life, and hence replacement of these was essential. The launch of the GSAT-5P satellite, possible with thirty-six transponders, was crucial for us. Also, four more GSATs were in the queue for launch by GSLVs in the coming years.

We decided on two courses of action: (a) launch the GSLV-F06, using a Russian cryogenic stage (delivered to us in 2004–06) to orbit the GSAT-5P communication satellite; and (b) try to procure a few more cryogenic stages from Russia. Also, I placed S. Ramakrishnan at the helm of the Liquid Propulsion Systems Centre from June 2010 after the superannuation of the director, M.K.G. Nair. I issued that order one month in advance for a seamless transition.

But further troubles were brewing 36,000 kilometres above the equator. In the wee hours of 8 July 2010, Dr Alex called. In an anxious voice, he said, 'We have a crisis regarding the INSAT-4B. Power supply from one of the two solar panels has been badly affected. Our mission team at Hassan saved the INSAT-4B with some swift action, but we might be left with only half the transponder capacity.' The INSAT-4B communication satellite had twenty-four transponders (twelve in Ku-band and twelve in

C-band), and was in operation since March 2007. A relatively young satellite, INSAT-4B transponders were the backbone of direct-to-home service providers such as Sun Direct, besides hosting VSNL and several small television channel operators. The fault was far-reaching.

We arranged leased transponders from foreign operators. The root cause of the snag was an electric arc in the slip ring (through which power collected by that solar panel gets transferred to the power management unit of INSAT-4B). A similar unit used in a W2M satellite, built by ISRO in 2008, had faced the same issue. We had a serious review. If a similar arc developed in the second solar panel's chain, it could end Doordarshan's DTH service. Also, if a similar flaw occurred in an identical pre-runner, INSAT-4A, Tata Sky would be in danger. We suggested that they operate their DTH service at lower power levels even at the cost of signal loss during heavy rains. This was indeed a relative disadvantage vis-à-vis their competitors in a highly competitive marketplace. This became an apt cause for sustained irritation with DOS.

The PSLV-C15 too was being assembled at the other launch pad in SHAR. This was to be the first launch after GSLV-D3. Our target was to launch PSLV-C15 by the second week of May 2010, i.e. three weeks after GSLV-D3. The PSLV-C15 launcher had been assembled by the third week of March 2010 itself and checkout operations were in progress. During the leak checks for the liquid stages, a marginal pressure drop was noticed in the second stage of the launcher. Since it was only a marginal pressure drop, it could be ignored if the launch was to take place within a couple of days. M. Annamalai, former director of SDSC-SHAR objected, and rightly so. Finally, his wisdom prevailed and the decision was taken to observe the pressure drop for a period of three weeks. Annamalai had been one of my senior advisers at VSSC and at Antariksh Bhavan. As we discussed this issue, I handed him a yellow slip and told him to use it if required during the experts' meet. 'My priority is to ensure a successful flight rather than an early one which might fail,' I wrote.

It turned out that the pressure drop was indicative of a major snag. PSLV-C15 was de-stacked for closer examination of the second stage in a test facility at Sriharikota. It was found that the pressure drop was caused by metal particles stuck on the base of an oxidizer valve. If any such particle, trapped in the oxidizer line, entered the Vikas engine during the flight, it would lead to a disastrous failure. I participated in a meeting held at Thiruvananthapuram. We decided to bring the second stage of the PSLV-C15 back to Mahendragiri for a detailed check-up. Every expert's viewpoint was heard. I was of the opinion that we should replace the Vikas engine, even if it meant delaying the mission by a month. Some felt that I was being overprotective and conservative, but I stood my ground. It was more important to think about long-term repercussions rather than give up for the fear of being called conservative. At that point, we could not have given a green signal without being 100 per cent sure. Our efforts paid off when PSLV-C15 flawlessly delivered the Cartosat-2B satellite into the intended polar sun-synchronous orbits on 12 July 2010. Three foreign satellites and a tiny satellite built by engineering students were also launched.

The presence of Prof. M.G.K. Menon, deputy chairman of the Planning Commission Mr Montek Singh Ahluwalia along with his wife and noted economist Dr Isher Ahluwalia, and member (science) of the Planning Commission Dr Rangan in the VVIP gallery of the Mission Control Centre encouraged us; their words were reassuring to the nation. After the successful injection was announced, I introduced the centre directors, mission director Kunhikrishnan and the team to the media and prompted them to speak. I stood behind.

Cartosat-2B had a critical postnatal period with a contaminated propulsion system that was revived with considerable effort. The mission wasn't affected, and it was successful. But the cardinal point was that within three months of a major failure, we had sprung back to execute another mission with success.

In July 2010, ISRO was granted an opportunity to brief
the PM on the successful mission. I led a team of about fifteen
functionaries to his residence. The centre directors graciously
stepped back and let their associate directors join me for the
briefing. I made it a point to brief Delhi on the glitches that we
had faced in the satellite propulsion system and the way we had
tackled it. By then, I could sense the PM's trust and confidence
in me. The regular briefing sessions at '7RCR' strengthened
our bond. And then the PM granted me the rare privilege to
call him up whenever I required. I used this entitlement with
due discretion.

My first global interaction as the space agency chief of India
came during the International Astronautical Congress of 2010,
held in Prague during the last week of September. I participated
in the Heads of Agency plenary on the opening day, along with
NASA administrator Charles F. Bolden Jr, head of Roscosmos
Anatoly Perminov, ESA director general Jean-Jacques Dordain
and JAXA chief Keiji Tachikawa. Incidentally, that was also
Bolden's first International Astronautical Congress (IAC) as the
NASA chief. We had a bilateral meeting and then continued to
interact through several other forums.

During the Heads of Agencies Summit, the respective chiefs
shared an overview of their current programmes, and gave an
insight into future plans and potential international opportunities.
After the interaction, the floor was thrown open to the audience
for a question-and-answer session. IAC being the largest
congregation of space engineers and professionals, the audience,
naturally, comprised the who's who of national space agencies
and the space industry from across the globe. As the president
of the International Academy of Astronautics, my predecessor
Mr Madhavan Nair was seated in the front row of the auditorium.
Recently, while flipping through an old photo album, I came
across pictures of V.K. Dadhwal, S. Somanath, P. Kunhikrishnan,
Surinder Singh and Santanu Chowdhury—delegates who had
attended the bilateral meets with other space agencies. They are

all directors of ISRO's major establishments now. The summit was a satisfying experience.

In November 2010, HYLAS, a communication satellite built by ISRO and EADS-Astrium of France, was launched on board Ariane-5 from Kourou in French Guiana. The mission was successful and the satellite was delivered to its final orbit for a UK customer—Avanti Space Ltd. That was a commercial contract earned by Antrix. This was lauded internationally as it added another feather in ISRO's cap.

ISRO's ability of developing state-of-the-art communication satellites was gaining global attention.

In India, the space system had created a robust space-based broadcasting and communication infrastructure. However, we were still struggling to keep up with the exponential demand.

No capacity had been added since INSAT-4CR was launched in 2007. Two satellites—EDUSAT and INSAT-3B—had retired in the meantime. To supplement our national capacity of 160 transponders, ISRO had leased nearly seventy-five transponders from international satellite operators like INTELSAT, SES, MEASAT, etc. It was a precarious situation with pressing national imperatives queuing up. We had to wait for another GSLV, with thirty-six transponders on board, to orbit our GSAT-5P communication satellite. Just two days after the launch of PSLV-C15, we started stacking the next GSLV (GSLV-F06) with a Russian cryogenic stage. By the second week of December 2010, the GSAT-5P satellite (2310 kilograms) was mated with the GSLV-F06 launcher at Sriharikota, and the integrated launcher and satellite were being checked for their readiness to fly. There were a few replacements and retestings during the final assembly phase; GSLV-F06 looked fine. I attended all the Mission Readiness Reviews to keep track of things. A joint team comprising engineers from ISRO and from the Russian supplier was mandated to review and clear the cryogenic stages in GSLV since its very first flight in 2001. This ISRO-GK Commission, after due consultation, certified the CS-6 cryogenic stage (one of the two latest stages delivered to ISRO in 2004–2006) to be assembled with GSLV-F06. These two stages were upgraded for more propellant loading. Hence, additional qualification testing was required. I had a special session with this joint team, particularly focusing on the additional qualification tests conducted in Russia.

The Russians were satisfied that all requirements specified for preservation were met at Sriharikota. They certified that the CS-6 stage was in good health for flying. Everything looked all right. The launcher was cleared for movement from its assembly bay to the launch tower. Kalam Sir made a visit to Sriharikota, witnessed the launch preparations, mingled with the teams and sat through a brief presentation. This brought cheer to the team. Kalam Sir always brought a lot of positivity wherever he went.

The Second Strike

The launch of the GSLV-F06 was initially scheduled for 20 December. The prelaunch days were eventful. Just a day prior to the countdown, a leak was identified in the thermostating valve of the liquid oxygen tank at a rate higher than the prescribed limit. Also, we had doubts if the leak would be aggravated by the vibration experienced during the initial period of the flight. We called for an emergency meeting of the Launch Authorization Board, which recommended a specific experiment to be conducted along with the resident Russian team to ascertain the stability of the leak while subjecting the valve to cryogenic temperature. Thankfully, the experiment proved that the leak rate was stable. Analysis indicated that adequate margins were on board to tackle this leak and hence clearance was given for the launch operations.

When we plan a launch from the second launch pad at SDSC-SHAR, the rocket is built in the adjacent Vehicle Assembly Building (a mammoth eighty-metre-tall structure) on the launch pedestal (or 800-tonne mass). About two weeks before the launch, the rocket mounted on the launch pedestal is moved to the launch pad, one kilometre away, on rail tracks. The rocket is connected to the umbilical tower on the launch pad thereafter, for necessary services such as fuelling, electrical checking, etc.

As luck would have it, there was a torrential downpour a few days after we had rolled the GSLV-F06 launcher to the launch pad. It was exposed to the rain and there was water ingression in one of the electronic packages of the cryogenic stage. We replaced all critical electronics and repeated the electrical check. Additional rain-proofing of the suspected areas was also done. Finally, after the health of the launcher was confirmed, clearance was given for commencement of the countdown on 24 December 2010 for a launch on 25 December 2010. I reached SDSC-SHAR two days in advance after our customary visit to the Tirumala temple. I also participated in the special Christmas mass that was arranged in the forenoon at a tiny Church at Sriharikota.

Mr Madhavan Nair and other veterans of ISRO were present at Sriharikota to witness the launch. However, our beloved Pant Sahib, who had been chairing the Mission Readiness Review Committees for a couple of decades, had been hospitalized a month ago, just before one such review of the GSLV-F06, on 18 November 2010.

The thirty-hour countdown went off smoothly. The GSLV-F06 lifted off as planned at 1604 hours (IST). All four liquid strap-on stages (L40s) ignited 4.8 seconds prior to lift-off, as required. After the normal chamber pressure was ascertained in all four strap-on engines, the mission computer gave a green signal for ignition of the core solid stage (S139 stage). The lift-off was majestic. The range operations director soon announced the normal operation of the core solid stage (S139) and the liquid strap-on motors (L40s). The launcher steadily gained velocity and altitude. On one of the screens, I saw the formation of the beautiful condensation ring around the heat shield of the GSLV. Formation of the condensation ring is quite normal when the rocket passes through the humid atmosphere at the velocity of sound. Suddenly, I heard Dr Alex exclaiming, 'What's happening there!' He was looking at the visual pictures from high-speed cameras. I noticed that the chamber pressures of the S139 core stage and four strap-ons had dropped to zero, and almost at the same time we heard the range safety officer announcing, 'Command issued.' He was referring to the 'destruct' command that is issued from the ground station to the launcher when the rocket loses course and poses a threat to the range itself as well as the habitat and installations en route. As the flight takes place, the range safety officer and his colleagues keep a close eye on the vital statistics of the rocket and also keep analysing the flight data. In case of an anomalous flight, they are authorized and mandated to issue the 'destruct' command to the rocket. As GSLV-F06 appeared to be faltering dangerously, they had to issue the command. The fragments of the time/altitude curve were showing dotted lines of different colours, depicting the data from several sensors that were mounted on different parts of the rocket. The GSLV-F06 mission too ended as a failure.

It was a ghastly sight as five disintegrated and burning rocket stages of the GSLV-F06 somersaulted into the Bay of Bengal. Unfortunately and ignorantly, the visual media described it as a 'mid-air explosion', possibly taking their cue from the multiple colours of the flame that appeared because of the different types of propellant in the rocket stages. Later, they found an even worse moniker—the 'epic failure of GSLV'. It was painful treatment. That multicoloured spiral pattern was called 'Radhakrishnan's signature' by the media for a couple of years.

Again it was a tough situation for me to announce the failure, and sadly enough this was the second one within a gap of eight months. Our immediate priority was to oversee emergency operations. Luckily, everything was fine. There was no damage to the range as the GSLV-F06 had been far enough from the shores of SDSC-SHAR. Immediately, we had a quick look at the radar data, video footage and flight data. In short, whatever was available in that short span of time. The performance of the launcher had been normal up to 47.5 seconds from lift-off. The events leading to the failure got initiated at 47.8 seconds. From the readily available data, we could decipher that at that stage the launcher had started developing bigger errors in its orientation, leading to pressure on its structure. The launcher could not sustain under this load and consequently it started disintegrating at 53.8 seconds from lift-off. The 'destruct' command was issued from the ground at 64 seconds. The flight was hence terminated in the first stage itself.

That day, the PM was on a visit to Kerala. I sent an SMS to his private secretary, Jaideep Sarkar. Later I learnt that by then the PM had already boarded the flight. After landing in Delhi, he called up. As I explained what had happened, he consoled me saying, 'These things happen. Don't get discouraged.' Those words gave me courage and strength. Kalam Sir, Prof. Rao and Dr Rangan also telephoned and offered their help.

The scene of April 2010 repeated itself eight months later. As the head of the organization, I took sole responsibility for the failure before the national media at the Brahm Prakash auditorium. After explaining our quick observations, I promised to come back with more data within a couple of days after our teams came to grips with the incident.

I requested Mr Madhavan Nair to join me during the preliminary review held at the Kalpana guest house along with the Launch Authorization Board, Mission Readiness Review team, Flight Readiness Review team and the GSLV mission executives. This hall had been a witness to several such post-failure scenes, but none of them had been more impactful than this. The flight data was quickly synthesized and after a quick analysis, we found that that there had been inadvertent snapping of a group of ten connectors mounted on a shroud at the bottom of the cryogenic stage. Some of these connectors carried command signals from the computer residing in the equipment bay, located near the top of the launcher, to control the four L40 strap-ons of the first stage. These connectors were intended to separate only on the issue of a separation command at 292 seconds after lift-off. The premature snapping of these connectors had led to the stoppage of continuous flow of control commands to the first stage. Consequently, this had led to the loss of control, which caused the development of errors in attitude control and the eventual disintegration of the rocket. That much was clear to us by then.

The exact cause for the snapping of the connectors was still to be pinpointed. We still didn't know if it was due to external forces like vibration, dynamic pressure or something else.

We returned to Bengaluru the next day via Tirupati, after a darshan at the temple. Madhusudhana walked into my office at Antariksh Bhavan. I greeted him as he came to my table. He took one look at my face and said, 'I just wanted to see how you have taken this failure. I appreciate your steely nerves.'

I immersed myself in finalizing a four-pronged action plan for recovery. Firstly, a thorough failure analysis of the GSLV-F06 and a closer look at the flight data of all previous flights was needed, and I requested Mr Madhavan Nair to chair this committee, along with a few experts from ISRO and outside. Secondly, we needed a programmatic review and strategy to recover from the impact of the GSLV failures, specifically to meet the demands of communication transponders; for assured launches of the satellites that were to fly on board GSLV in the next few years. These included the INSAT/GSAT series of communication satellites, the INSAT-3D meteorological satellite and Chandrayaan-2. We also had to focus on the development of the Indian cryogenic stage. Dr Rangan was most suited for this task, especially with his national role as member (science), Planning Commission.

Internally, we needed to introspect and I wanted the guidance of Dr S.C. Gupta, a former member of the Space Commission and former director of VSSC.

Finally, I sought the collective wisdom of a group of 'elders'— Kalam Sir, Prof. M.G.K. Menon, Prof. Yash Pal, Prof. Rao, Dr Rangan, Mr Madhavan Nair and two members of the Space Commission, Dr R. Chidambaram and Prof. R. Narasimha.

I set up these four mechanisms on 29 December 2010. Before signing the orders, I wrote a short briefing note to the PM on the

action plan and his endorsement was conveyed over the phone by
the national security adviser, Shivshankar Menon.

The year 2010 was a tough one for ISRO and keeping up the
morale of the organization was our top priority. My New Year's
address to the entire ISRO community on video was a call for
introspection and to move ahead. I said:

> Today, we are stepping into 2011 with a bed of fire in front
> of us, to walk through unscathed and to come out with flying
> colours . . . We must not fear failure. We should embrace it
> but also learn from it. Every dark cloud has a silver lining.
> Admitting and learning from failure will ultimately lead to
> success . . . . It is important to reassure the nation of the process
> of the review and analysis of what happened; the steps to be
> taken at this crucial juncture of the Indian space programme;
> and give adequate visibility to the entire process. This will be a
> full-fledged review and we cannot leave any stone unturned . . .
> Friends, this is a time for serious introspection. I am going
> through a comprehensive exercise of introspection within ISRO
> to retune the organizational structure, systems and processes.

Clearing the Air

Within a fortnight, I conducted one-to-one sessions in an informal
setting at my home office, with thirty senior scientists (including
seconds-in-command in all ISRO centres, programme directors,
project directors) to capture their feedback on a structured
questionnaire that addressed programmatic issues, priorities,
efficacy of our current structure, systems, review mechanisms, etc.
Also, I sought their frank feedback on my methods of running
ISRO, and I got some insightful and constructive suggestions. I
encouraged them to speak about their aspirations and the way
they could contribute more to the organization.

As if the agony from the failures was not enough, in a couple
of days I received a scathing email from a perceived well-wisher

and benefactor. The message carried several baseless allegations, even going to the extent of saying, 'Now you have nullified the whole process, the centre directors and senior colleagues, in essence the whole ISRO community is decimated.' In the same email, there were several harsh statements about the veterans roped in for the review exercise. It was unimaginable if not unethical for someone who had been in the thick of launch failures many times to shoot off an email like that. This was unbecoming of a fellow professional. I preserved that harangue in personal archives.

To top it all, I was told there had been complaints against me in Delhi citing that I had not been able to work in tandem with my senior colleagues and this was the reason for the back-to-back failures of the GSLVs in my regime. I made it clear to Delhi that I did not wish to continue as the chief of ISRO, if Delhi did not have confidence in me and if my colleagues were not with me.

I requested the scientific secretary, V.S. Hegde, and the new additional secretary, R.G. Nadadur, to talk and seek feedback from each of the centre directors and convey the same to the PMO directly without keeping me in the loop. The centre directors chose to collectively meet me to express their solidarity. The decision makers at Delhi were quick to gather facts about the previously failed and erratic GSLVs and comprehend the situation. It was a tough time, but I held my ground.

I had kept myself fully detached from the proceedings of the Failure Analysis Committee of GSLV-F06, as I wanted it to be free from the perceived influence of the chairman.

The Failure Analysis Report arrived on 2 April 2011. The main conclusions of their three-month-long analysis were: (a) the primary cause of the failure was the excessive static deformation of the shroud due to aerodynamic loads, which resulted in untimely snapping of all connectors between the second stage and the cryogenic upper stage, and (b) that the above led to stoppage of flow of command signals to the first stage control power plants from the equipment bay, leading to the vehicle losing control.

Further, they stated that the failure had occurred due to a combination of factors such as (i) design inadequacy on the mounting scheme of these connectors on the shroud, (ii) not having an exact assessment of the local aerodynamic loads acting on the shroud, and (iii) possible deficiencies in the mounting of the connectors on the shroud and the interface fastening of the shroud with the cryogenic stage. Also, they pointed out that there was a de-mating of one such connector in the GSLV-F04 flight (of September 2007) possibly due to the deformation of the composite shroud and the bracket to which the connector lanyards are anchored.

Ironically, the prime revelation was that such aerodynamic loads were present in all the past flights of GSLV. But, when it repeated in GSLV-F06, the result was catastrophic.

The question that stared us in the face was: Why did it become a catastrophe in GSLV-F06? Here, we had flown one of the two Russian cryogenic stages supplied to ISRO in 2004–06. A joint investigation on the remaining and identical Russian cryogenic stages indicated that the slackness provided in it for such connector cables was far below the specifications—poor workmanship that had gone unnoticed at all stages of inspections there. The catastrophe could have happened in the case of GSLV-F04 as well, but it actually manifested only during the GSLV-F06 flight. Both sides learnt it this time, but it was a costly learning process.

Dr Rangan's committee on programme recovery ensured that my views were taken on board. This committee made valuable recommendations which included:

(a) Priority for launch of GSLV using Indian cryogenic stage and support development of LVM3
(b) Launch of INSAT-3D and GSAT-7 from abroad
(c) Build 1000-kilogram-class communication satellites and launch by PSLV till GSLV launches get stabilized

(d) Build and launch additional 2000–3000-kilogram-class communication satellites

(e) Lease satellite transponders from foreign satellite operators and even lease foreign satellites, to be operated from orbital slots allocated to us

(f) Buy or build heavier and high-power satellites.

The committee emphasized enhancement of quality and reliability culture within ISRO and its associate industries.

Distinguished elders—Kalam Sir, Prof. M.G.K. Menon, Prof. Yash Pal, and Prof. U.R. Rao—were briefed, and they endorsed the recommendations of these two committees; the Space Commission concurred and we finally moved ahead.

In the meantime, in February 2011, ISRO was battered again. I had to prove my integrity, inner strength and sense of discretion as well as communication skills. A national news agency reported an astronomical loss of Rs 2,00,000 crore through the Antrix-Devas agreement of January 2005. Of course such a sensational claim turned out to be false after authentic reports were submitted by authorities like the CAG, a year later. But the credentials of a nationally adored organization were under the public radar, besides the political and strategic dimensions of the case were too intricate. Steering ISRO through this conundrum was a tough task. I had to face the music for something that happened when I was not even with the Department of Space. The Antrix-Devas fiasco continued like a wild background score with several spurts at irregular intervals, but I learnt to live with it.

But the morale of the ISRO family was a matter of concern, especially since some of our past colleagues were involved in the fiasco. As the leader of the pack, it was my duty to distinguish between the chalk and the cheese and communicate the issues at hand. I went around all centres of ISRO and explained the myth and reality of this controversy. Transparency prevailed; I was touched when the LPSC family at Valiamala in Thiruvananthapuram gave me a standing ovation at the close of a one-hour talk.

With a deep concern for this aberration at ISRO and to bring back the morale of the teams, the P.M showed the rare gesture of coming down to ISRO's Space Applications Centre at Ahmedabad. He visited a few laboratories, interacted and mingled with the employees and then made a passionate address to the entire organization, which was broadcast live to all centres. The governor and chief minister of Gujarat also joined him for that session on 26 March 2011.

22

THE TURNAROUND

The freezing winter faded away, welcoming the blooming springtime in Bengaluru. The scorn and contempt for presiding over twin failures of the GSLV and the social condemnation for scripting the so-called 'S-band Spectrum scandal' receded when several institutions investigated and dug up facts. The factually correct and meticulous responses from both Bengaluru and Delhi also helped. The PM ensured that the morale of ISRO would not be affected; his confidence in my leadership was explicit through words and deeds.

This trying time did take a toll on my mental bearing. I turned to classical music for the healing touch and to recoup. There were days when I would wake up in the wee hours and start practising vocal music vigorously. I never bothered if I sounded great or not; the very exercise gave me tremendous amounts of mental calmness to endure what those excruciating days had to offer. Music and prayer closely intertwined and they kept me going (most of the Carnatic music I practise is devotional). It became a wonderful regenerative process.

Week-long visits to Delhi to attend difficult meetings became more frequent. I was required to brief different authorities on the same subject repeatedly. And then the cycle would repeat itself the

201

very next week. It was traumatic, but I braved it. At the same time, we also had to make sure that we did not programmatically stop. At ISRO, we needed to forget the past and immerse ourselves in the imminent missions at hand.

Focusing on the Tasks at Hand

The Resourcesat-1 satellite, launched in October 2003, had the pivotal role of India's remote sensing applications to forecast agricultural crop production, investigate water resources, create an inventory of forest resources, monitor land use and rapid survey of areas under disaster distress. Several users from India and customers from abroad were dependent on the data it produced. By that time Resourcesat-1 had outlived its in-orbit assured life of five years. Though we kept using it for a couple of years more, the service was crippled. Finally, we were ready with Resourcesat-2, with a few new features, to be launched on board PSLV. We flagged off the satellite from the ISRO Satellite Centre, in the last week of December 2010, within three days of the GSLV-F06 failure.

A sense of extra caution was mandated at the launch base, Sriharikota, during the mission campaign of PSLV-C16 which was to loft Resourcesat-2 and two co-passenger satellites. The co-passenger satellites were of much smaller sizes—Youthsat, an Indian satellite (with one of the payloads from Russia), and a Singaporean experimental satellite called 'X-Sat'. Based on the learning from previous missions, PSLV-C16 had a few technical issues that had to be resolved; this was meticulously executed by the project team. A flawless launch by PSLV-C16 was ISRO's lifeline at that juncture.

As a mark of solidarity, the PM deputed the MOS (PMO), V. Narayanasamy, to Sriharikota. Prof. M.G.K. Menon and Dr Rangan also joined us on my invitation. The launch was successful. While mission director Kunhikrishnan's interview was being aired on national television, I got a call from the PM, congratulating the ISRO team. Resourcesat-2 functioned perfectly. Within a fortnight, the PM met the ISRO team at his official residence. This time, owing to the gap, I was able to present to him a few astounding imageries from the Resourcesat-2. ISRO got a reprieve. However, a section of the media continued its tirade on the Antrix-Devas issue!

The successful model of international cooperation on the Chandrayaan-1 mission (it hosted six foreign payloads) was reaping dividends. A lovely invitation came to us from NASA's Jet Propulsion Laboratory (JPL) to team up for a Lunar Sample Return Mission, initially titled 'Moon Rise'. As I said earlier, nothing excited ISRO more than a challenging mission. The proposition of working with JPL again, and that too on such a prestigious mission, truly acted as another adrenaline booster. Though the 'Moon Rise' mission, one of the three potential planetary missions considered for NASA's 'Medium Class New Frontier Probe', did not pass the final evaluation, it marked the beginning of a new phase of intense India-US space cooperation.

By March 2011, we were ready with the GSAT-8 communication satellite. GSAT-8 hosted twenty-four Ku-band communication transponders and this payload consumed nearly 6250 watts of electric power, the highest that ISRO had handled till then. Also, GSAT-8 was carrying the first GAGAN payload. A previous attempt at launching GAGAN on board GSAT-4 did not materialize due to the failure of GSLV-D3. Effectively, the GAGAN payload of GSAT-8 became the first of the three to orbit on the GSAT series of satellites. GSAT-8 weighed 3200 kilograms, almost 1000 kilograms more than what a GSLV could loft. Hence the launch was done with a European launcher, Ariane-5, from Kourou, French Guiana. The prelaunch preparations of GSAT-8, spanning forty-five days at Kourou, were ably handled by the duo—Prahlad Rao and Venkat Rao—with a team of ISRO engineers. Dr Alex and I joined them a few days before the scheduled launch.

The launch base at French Guiana on the north Atlantic coast of South America is quite close to the equator, and hence inherently advantageous for placing geostationary satellites at the circular orbit of 36,000 kilometres above the equator. The uninhabited territory to the east makes the launch easier as it reduces the risk of human casualties in case of a failure and breaking of the lower stages of the launcher. But the campaign was far from the expected cosy experience of a foreign location. The only mode of transportation to this tiny South American base was a transatlantic flight from Paris that took nine hours. The general visitors were either the Arianespace employees, their customers or French government officials who took care of the colony. For the ISRO team, the first challenge was to carry the huge satellite in a cargo aircraft from Bengaluru to Kourou. This was almost like carrying a newborn baby in an incubator. Second, the launch teams had to go fully equipped for the prelaunch assembly, tests and medical emergencies. The team was accommodated in tiny dwellings (the only ones available) during their long stay in the hot and humid conditions of the Amazon rain forests. To top it all, Kourou is a nightmare for vegetarians, and 80 per cent of our team followed that diet. So, essential food items like rice, lentils and masalas had to be carried, and our team members had to cook their own food during the stay. Life at our own launch base on the secluded island of Sriharikota was easy in comparison.

That was the fifty-eighth launch of the Ariane-5 launcher and the 202nd launch of the Ariane family of launchers, which had its first launch in 1979 with the Ariane-1 series. There was a well-oiled system in place at Kourou, French Guiana, for Ariane launches. Still it was rocket science at play; pending reconfirmations of certain observations mandated for the launcher had not been cleared. Inclement weather threatened further delay. GSAT-8 was in good shape. The launch finally happened on 21 May 2011, and it was a success. Our team at Hassan took charge once the Ariane-5 placed GSAT-8 in the geostationary transfer orbit. The

PM called me up at Kourou to congratulate the team. Along with Dr Alex, I joined the project team for a 'celebratory lunch' at their temporary dwellings. The menu comprised an authentic Karnataka sambhar, rice, avial (a mixed vegetable curry) and curd; this was one of the most satisfying meals I had ever had.

Within a month, GSAT-8 was placed in its assigned slot in the geostationary orbit; twenty-four Ku-band high-power communication transponders were added to our national capacity. ISRO, along with the Airports Authority of India, started testing the 'signal-in-space' from the GAGAN payload.

Our next mission was to orbit a communication satellite, GSAT-12, with PSLV. To improve the capacity of communication transponders, ISRO devised a clever strategy in the year 2009, to employ a high-end PSLV to launch GSAT-12 into an elliptical orbit as high as it could, and then raise both its apogee and perigee by five steps to place it in the final geostationary orbit. On 15 July 2011, PSLV-C17 completed the job. Raising the apogee was a novel operation for ISRO at that point of time but our teams carried it out precisely. I invited the national media to join us at Hassan, and an amazingly large and senior team sat through with us as the final phase of antenna deployment operations was executed.

I addressed the entire ISRO community from Hassan and prefaced my talk with the punchline, 'ISRO is back on track'. That extempore speech was from the bottom of my heart, complimenting the teams for the turn around. For me, there were three principal success factors behind this resurgence—(i) the primacy granted to reliable engineering teams; (ii) inculcation of cautiousness in technical decision makers; and (iii) a democratic approach to resolve productive conflicts between managerial decision-making forums and the established review mechanisms. All these worked perfectly in sync.

We tweaked the traditional preflight review mechanisms; clearances from Flight Readiness Review (sub system level) and Mission Readiness Review (whole-system level) were mandated even before the assembly of the launcher commenced. This avoided the predicament of deciding whether to de-stack a fully assembled launcher when red flags were raised in reviews just prior to a flight.

ISRO's early failures in the 1970s and 1980s could be attributed to our lack of understanding of the behaviour of launchers and satellites in certain space environments. Many of the subsequent failures (including those we faced during the past four launches of the GSLV) could be traced to the malfunctioning of one or two components. At times, a non-conformance to any of the specifications would escape scrutiny during manufacturing, assembly or preservation, in spite of multiple tiers of checkpoints. Sometimes, a software bug or a failure mode went unnoticed as it did not create any issues during the flight. We had to catch them at the point of origin itself. We initiated a directorate of system reliability and quality as a watchdog at the headquarters with the competent S. Selvaraju, with four decades of domain expertise at VSSC, as its chief. At all ISRO centres we tightened the inspection, quality control, quality assurance and quality audit systems with protocols and procedures.

That was essential but not sufficient to ensure successful missions at an enhanced rate for a longer run. It is well known that any building is as strong as its foundation. We had about 5000 colleagues, at different levels, who machined, inspected, assembled and tested the articles to realize a rocket or a satellite. In essence, they had to be brought on board about the flaws that could creep in at that level. Such minor flaws could cause a launch failure or malfunction of a satellite already in orbit. With that thought we initiated the 'zero defect delivery system' in 2012. We roped in our distinguished professors, Dr B.N. Suresh, Dr R.R. Navalgund, Dr T.K. Alex, M. Annamalai and S. Selvaraju, to take this mission across all centres and units. Mostly, they resorted to workshops of small groups at the workplace (system delivery agencies); deliberations were conducted in local languages as much as possible. One of the key elements was systematic documentation of the procedures to transfer the know-how to the younger generation.

With the GSAT-12, India's space-based assets rose to nine geostationary satellites (INSAT and GSAT series) and eight polar orbiting satellites. Ground control for upkeep of this satellite fleet was a demanding task for the mission controllers at ISRO's Master Control Facility (at Hassan and Bhopal), and at the ISRO Telemetry Tracking and Command Centre (at Peenya, Bengaluru). Space applications, being the mainstay of the Indian space programme, have a very high commercial potential with a high-stake downstream market. This holds good for communications and broadcasting services in several sectors as well as satellite-based remote sensing data and services. In India, the stated demand for communication transponders had surpassed the supply significantly. We moved forward with caution and the collective wisdom of the revived INSAT coordination committee of secretaries to the Government of India, as the prevailing scene on allocation of precious national resources was complicated at that time. Meanwhile, the Remote Sensing Data Policy, revised by the government in July 2011, eased restrictions on disseminating satellite data. Under the new policy, ISRO was authorized to disseminate satellite data with resolutions up to one metre on a non-discriminatory basis, whereas data with spatial resolutions below one metre could be disseminated following a prescribed process and procedure. That meant, in practical terms, that data from Cartosat-1 and Cartosat-2 could be made available to a wider user community.

The national planning committee of secretaries to the Government of India, chaired by Dr Rangan in his capacity as member (science) Planning Commission, gave impetus to the institutional mechanisms. A national centre in the ministry of agriculture for crop forecasting and monitoring agricultural drought, an entity within the ministry of water resources for national water resources information systems, facilities at INCOIS to receive data directly from Oceansat-2, a national information system for climate and environment research at the Shadnagar campus of NRSC, etc. were commissioned soon.

A major organizational change came in July 2011. The selection-cum-search committee (of which I was a part) recommended the appointment of Dr V.S. Hegde as full-time chairman and managing director of Antrix, and the recommendation was approved by the government. Antrix got the autonomy it deserved. I was happy to monitor the progress of Antrix in my capacity as secretary of the parent department. We implemented a set of governance and systemic reforms within Antrix, Department of Space, and in the conduct of business in the Space Commission, with defined goals and rules of engagement.

The Right People

Now, we needed a new scientific secretary to head the headquarters. Besides being a programmatic head, he/she was required to have a broad understanding of the space programme and a flair for international relationships. An ability to communicate with Delhi was an added advantage and, above all, the candidate was required to complement my strengths and judgements. We had a few candidates in mind but my first choice was Koteswara Rao, who was heading LEOS at the time. I had been thoroughly impressed by his technical acumen which was evident in many reviews of our satellite programme. His balanced approach to techno-managerial issues and ability to communicate with precision and brevity were remarkable. When I mentioned him to the additional secretary, Dr Nadadur, he did not hesitate to endorse the choice. I discussed this proposal with Koteswara Rao at my home office. He told me about certain skill gaps that he felt could affect his performance in the new role. I commented that this realization in itself was commendable and a step towards success. Second, he wanted time to consult his mentor, Dr Alex. The next day, he expressed his willingness to move to the headquarters. At the end of his tenure, Koteswara Rao turned out to be one of the best scientific secretaries ISRO had ever had. I was a bit relieved to have him on board.

During my first briefing to the PM as the chairman of ISRO in November 2009, he had advised me to take advanced actions to groom the future leaders. That was in perfect sync with my beliefs. The current generation of space scientists and engineers at ISRO could be divided into three sections. Its first category comprised employees who had been inducted in the 1970s. They had grown with the organization and had worked on projects such as Aryabhata, SITE and SLV-3. The members of that generation were the current leaders of ISRO, to superannuate within a few years. The second category comprised engineers in the age group of thirty-five–fifty-five years who had been witness to the development of PSLV, GSLV, IRS and INSAT projects. This group comprised a large number of inductees of the 1980s and early 1990s. And then there was the generation of the 1990s and later—a younger lot comprising nearly 35 per cent of ISRO's scientists and engineers. The second category, in their flowering mid-career, was our target to bring out the future leaders. We initiated a thorough process in December 2009 and created the 'ISRO Strategy Group' of 225 high performers across centres and units of ISRO. I also roped in Mr Y.S. Rajan, the live wire scientific secretary to Prof. Dhawan and my mentor at the headquarters, to drive this group towards identifying the areas which ISRO had to concentrate on to create a vision document for the next twenty-five years. The aim was to absorb the global developments and sociopolitical implications in regard to the Indian space programme. And they did a fantastic job. The group essentially captured the length and breadth of the technological domains of ISRO as well as the different age groups.

When the PM visited the Space Applications Centre, Ahmedabad, in March 2011, the selected members were introduced to him.

A couple of months later, national security adviser Shivshankar Menon conveyed that the PM was keen to have a presentation on succession planning and leadership development for the next fifteen years at ISRO. The challenge was to delineate their age profile, residency and seniority in the current grade, track record, scientific, technical and managerial skills; potential for higher levels of responsibilities and broadening domains. The entire exercise was to be done with the utmost discretion for obvious reasons. A few of ISRO's founders, in their seventies, were still devoted to the space programme, and their wisdom was valuable to us in this exercise too. After data collection and discreet consultations with some of them, I immersed myself in the exercise, ably supported by the scientific secretary, Koteswara Rao, (who had crossed fifty-nine years and hence was not an aspiring candidate) and a young lieutenant of his. My cabin was declared 'out of bounds' for all except these two gentlemen for a fortnight. The outcome of this exercise was a 100-page document that described the desired profile for leadership positions in each of ISRO's centres/units and ideal candidates for two–three levels of transitions in each of them. The document also had a suggested succession plan for the post of chairman, ISRO, and secretary, DOS. I submitted the document to the PMO on 29 August 2011, precisely two years before the closure of my specified tenure (i.e. 28 August 2013). This seminal document was lauded by the PMO and it became a reference for subsequent seamless managerial transitions at ISRO centres.

There was a chain of changes in three major centres of ISRO. Kiran Kumar, Shivakumar and M.Y.S. Prasad moved up as centre directors in the Space Applications Centre (Ahmedabad), ISRO Satellite Centre (Bengaluru) and Satish Dhawan Space Centre (Sriharikota) in 2012. This was triggered by the superannuation of Dr Navalgund, Dr Alex at SAC and ISAC respectively and the transfer of Chandradathan from SDSC to LPSC. A relatively young Dadhwal had already become the director of NRSC.

We refined the upper strata of the ISRO strategy group. From a pool of 225, fifty most senior performers were guided through a grooming process for assuming immediate leadership positions at six centres and eight units. A select cross-disciplinary group among them was tasked to draft the Twelfth Five-Year Plan (2012–17) and defend it within the ISRO council, in front of a national-level steering committee and in a review by a member (science), Planning Commission. I vividly recall the brilliant presentations made by V.K. Dadhwal, K. Sivan, M. Annadurai, Tapan Misra, S. Somanath, V. Narayanan and V. Ashok at Delhi on a couple of these platforms. It is gratifying that the current leadership team, who are carrying the baton forward at all centres and units of ISRO, acknowledge that they benefited through this process.

In the middle of all this excitement, ISRO was on the threshold of another mission: an Indo-French joint satellite project 'Megha Tropiques' was ready to fly. The mission was conceived with CNES, the French National Space Agency, way back in 1999. As its title suggests, this satellite aimed to study the water cycle and energy exchanges in the tropical regions, a unique contribution towards climate research. I was acquainted with it while I was at INCOIS. Indeed it was a mission with great global relevance. On 12 October 2011, PSLV-C18 precisely orbited the Megha Tropiques satellite. Three tiny co-passengers—JUGNU of IIT, Kanpur, SRMSat of SRM University, Chennai, and Vesselsat-1 from Luxembourg—also orbited.

A Disturbing Background Score

Meanwhile, the frenetic activities around the so-called 'S-band Spectrum scandal' or the 'Antrix-Devas deal of January 2005' were taking away the focus from important tasks at hand, and affecting ISRO's institutional energy. Much against my expressed wishes, I had to sit as a member of a high-level team (HLT) chaired by the former chief vigilance commissioner, Pratyush Sinha. A report by the HLT was submitted to the government in the first week of September 2011. The government acted on this report without extending the matter further, and my trips to Delhi became more frequent. Finally, in the last week of December 2011, the government's considered decision, to exclude four retired ISRO scientists from any important roles under the government, was handed over to DOS. Also, it was directed to the state that these officers should be divested of any current assignment/consultancy with the government with immediate effect. This decision was communicated by DOS after several rounds of consultations with Delhi on the content, format and process.

On 25 January 2012, India woke up with a media uproar and brickbats flew again. I was on the receiving end of filthy personal attacks. It was painted as my personal agenda to deflect failures. I was compared with Judas Iscariot and Hitler. Many learned members of the scientific community jumped on the bandwagon blindly. The lone voice of reason was Kalam Sir's, who stated, 'Indian Space Research Organization is bigger than any individual. The organization will grow and succeed.'

Throughout, I preferred to keep quiet. I was part of a system and represented a reputed organization of the country; I could not afford to respond to all the allegations and drag the organization into a street fight. They say 'well done' is always one up from 'well said'. I chose to respond to the situation institutionally in tandem with Delhi, and placed the facts in the public domain by the second week of February 2012. I decided to concentrate on the job at hand. The media kept hounding me, and even mobbed me on many mornings at my residence, for 'just one byte' and scorned my *maun vrat*.

I had a clear conscience throughout this, I had done my duty in the national interest and duly consulted all concerned authorities. I formally sought advice even from my predecessor in June 2010 before commencement of the decision-making process at Delhi. At all crucial instances, I had done my best to help and not harm. I maintained that our focus was on missions and that successful space ventures would redeem the organization's pride. ISRO needed a public stance to face the raging onslaught collectively and it was imperative to reinforce confidence even within the organization, a section of which was sceptical about my intention. I decided to hold a two-day meeting of centre directors and senior functionaries at Sriharikota. We reaffirmed ISRO's commitments to the nation. We were engrossed, at that time, in the final assembly of PSLV-C19 due to launch India's first microwave remote sensing satellite, RISAT-1. Another team was assembling, for the first time, the giant LVM3 launcher for a series of trials. We had to take some serious corrective actions for the GSLV and the Indian cryogenic stage. In fact, the robustness and resilience of ISRO was wonderfully at work. But more importantly, the centre directors collectively addressed the entire organization from Sriharikota, with all establishments of ISRO and DOS connected on satellite link. We sought feedback on what more ISRO could have done last year; what more ISRO could do in the coming year; how to improve the processes in place; how to improve transparency; how to uphold ISRO's USPs: scientific and technical competence, focus, devotion and most importantly teamwork. We emphasized the fact that the success of the immediate missions should stand testimony to this. We reaffirmed to the nation that ISRO was fully focused on its programmes and nothing else really mattered.

In fact, ISRO had a head start in 2012 as President Pratibha Devisingh Patil came to Sriharikota on 2 January 2012. The President inaugurated the 'Mission Control Centre' and 'Launch Control Centre'; honoured our senior scientists and engineers; addressed the entire ISRO family via videoconferencing; and interacted with ISRO teams at the PSLV-C19 assembly hall and launch complex. PSLV-C19 successfully lofted India's first microwave remote sensing satellite RISAT-1 on 26 April 2012. At 1858 kilograms, RISAT-1 was the heaviest satellite ever lifted by PSLV.

The synthetic aperture radar of RISAT-1, operating in the C-band (5.35 gigahertz), was a technological marvel that only a few countries had achieved. The optical sensors in our remote sensing satellites relied on solar radiation either reflected or scattered from the objects on earth to decipher them; these sensors could not see through clouds. The synthetic aperture radar had its own source of microwave radiation to illuminate the objects on earth and a system to collect the microwave radiation back-scattered from these objects. Also, the microwave radiation could penetrate clouds. In simple terms, RISAT-1 was an 'all-weather' and 'day and night' eye in the sky, carrying its own 'flash bulb'. Though a complex process had to be followed on the ground to interpret this microwave remote sensing data, the resource managers were happy with this boon for periodic imaging of agricultural crops in the cloudy kharif season and surveying flood-inundated areas.

Incidentally, by May 2012, the comptroller and auditor general of India tabled its report titled, 'Compliance Audit on Hybrid Satellite Digital Multimedia Broadcasting Service . . .' in Parliament and soon released the report in the public domain. Several questions regarding the Antrix-Devas contract and the so-called 'S-band Spectrum scandal' were addressed.

In conclusion, within a span of one year, even in turbulence and thunderstorms, ISRO could clock ten successful missions—four diverse PSLV launches (PSLV-C16, C17, C18 and C19) and six very different satellites (Resourcesat-2, Youthsat, GSAT-12, Megha Tropiques, RISAT-1 and GSAT-10).

Truly, ISRO had come full circle and there was no looking back.

23

CRUISE CONTROL

The ten successful missions had given us some breathing space, and we now had the opportunity to assess the current scenario, immediate targets and expected outcomes for the coming years. We divided the future into three time frames—for the immediate future or the next three years we planned to focus on capacity augmentation in terms of space assets, infrastructure and throughput; in the medium term of up to six to eight years we decided to thrust towards capability enhancement through several ongoing initiatives for advancements including embracing international alliances; and in the longer run (next ten years) we wanted to scale the next level of the technology ladder, attempting missions that had never been thought about. Of course, establishing Indian presence in the solar system was important for this long-term vision. We also wanted to forge a strong Indian space industry to meet local and global demands. Together, with the centre directors and project directors, I made a fairly detailed presentation at the PMO and the Space Commission on this renewed assessment. Delhi was convinced that ISRO was on the right path towards the next level of excellence.

I had an immediate opportunity to share India's achievements in space on a global platform. The annual assembly of the

Committee on Space Research (COSPAR) was being hosted by India in July 2012, thanks to my predecessor Mr Madhavan Nair, who had secured it for us. Indian IT giant Infosys graciously offered their iconic and sprawling campus in Mysuru to conduct the event. Prof. Rao, as chief of the Local Organizing Committee, steered COSPAR 2012 with the same zeal with which he had delivered Aryabhata, the first space mission of India, in 1975. It was a congregation of 1800 space scientists from seventy countries, including several legendary seniors, who deliberated on the recent advances and future vision of all the domains of space science, including astronomy, astrophysics, atmospheric science, planetary science and microgravity.

Prime Minister Dr Manmohan Singh's message for the opening ceremony of 'COSPAR 2012' included:

Space is truly the final frontier for the human race. More than half a century since the dawn of the space age, space research and exploration continue to evoke wonder and awe. Space technologies and missions bring together the latest advances in multiple disciplines and their benefits extend to all humankind. Many of the instruments and services around us today that we take for granted would not have been possible without advances in space science. It is this perspective of using space research for the larger social benefit that led a developing country like India to embark on an ambitious space programme soon after Independence. Today, India has a robust programme that has helped us leapfrog technological developments and helped to reach significant social, economic and industrial advances to the remotest parts of our country. Fostering international cooperation has been a key element in these efforts.

Quite aptly, the pronouncement of India's first mission to Mars was done by the prime minister during his speech on the sixty-sixth Independence Day from the ramparts of the Red Fort. We had merely fourteen months to realize the dream of getting en route from the Red Fort to the red planet. We took on the challenge.

Milestones

In the meantime, ISRO got ready to launch its '100th space mission', on 9 September 2012. While India's maiden space mission, the Aryabhata satellite, was launched by the USSR in April 1975, its centum mission was to launch a French remote sensing satellite (SPOT-6) on board the PSLV-C21 launcher. It also had a co-passenger satellite—PROITERES of Japan.

The PM accepted our invitation to come down to Sriharikota on the previous afternoon. The visit made him nostalgic, as he had visited the island in its formative stage as member (finance), Space Commission. The majestic sight of the PSLV-C21, ready for lift-off, enthralled him. Flanked by the triumvirate, P.S. Veeraraghavan, S. Ramakrishnan and M. Chandradathan, the mission director, Kunhikrishnan, briefed him on the PSLV and the preflight operations conducted during the countdown phase. He was then taken to the adjacent Vehicle Assembly Building where the giant LVM3 was being fully assembled as part of a trial. The PM took a tour, during which M.Y.S. Prasad, K. Sivan, S. Somanath and V. Narayanan gave a presentation on the developmental tests lined up for the LVM3 as well as corrective measures for the GSLV and its cryogenic stage. The PM got a clear picture of all three launchers. Besides, he was happy to interact with three generations of rocket engineers of ISRO. The MOS (PMO), V. Narayanasamy, and the national security adviser, Shivshankar Menon, (both members of the Space Commission) mingled with a team of relatively young engineers at the Launch Control Centre and Mission Control Centre; they were navigating through the propellant filling operations and monitoring the health of the launcher through a few hundred parameters.

After the visits, in the privacy of his suite, I briefed the PM on the prerequisites for final clearances of the launch and alerted him that any extreme eventualities in terms of the hardware or weather may lead to postponement of the launch. He smiled and gave his consent. I sought his permission to be on my hot seat at the mission console and proposed that he be briefed by Dr B.N. Suresh at the VVIP gallery during the launch. Dr Suresh was the chairman of the Mission Readiness Review Committee, and I could not think of anybody more appropriate for this job. The PM happily agreed to this also. The next morning, the launch of PSLV-C21 was a grand success and a beaming PM gave an inspiring address from the Mission Control Centre. He passionately recalled his past association with the space department and with Prof. Satish Dhawan. On the way out, he shook hands with the mission executives, and a few young engineers, who were chosen to be there.

Within twenty days, we launched the GSAT-10 communication satellite, carrying thirty transponders, on board Ariane-5, from the European launch base at French Guiana. This time, I broke a four-decade-old convention. Till then, the chairman of ISRO used to be present at the launch base at Kourou with the satellite team during the last days of the launch campaign. I too had gone to Kourou in May 2011 for the launch of the GSAT-8 communication satellite, but this time I felt that I should be present with our mission team at the Master Control Facility (MCF), Hassan. This team performed critical operations after the satellite was injected by the foreign rocket. GSAT-10 was successfully launched and our mission team did their job flawlessly to put GSAT-10 in the operational mode. From then on, I made it a practice to be stationed at Hassan during our satellite launches from Kourou.

Within six months, Sriharikota was privileged to host the President of India, Pranab Mukherjee, to witness the launch of PSLV-C20 with SARAL or Satellite with ARGOS and ALTIKA. SARAL was ISRO's second collaborative satellite mission with the French national space agency, CNES. Its aim was to study the ocean surface and facilitate data collection from in situ platforms. Six co-passenger satellites from Austria, Canada, Denmark and the UK also rode to space along with the SARAL. The President arrived a few hours before the scheduled launch, on the afternoon of 25 February 2013, and drove down to a safe viewing point of the launch pad. The presence of the French ambassador and foreign customers at the VVIP gallery symbolized the international cooperation and commerce at work. After the successful launch, a jubilant President gave a scholarly address and interacted with the mission executives. He said to me, 'You must be the most relieved person on planet earth right now.' It reflected how he felt during the launch. Since then, I made it a point to alert the President's office of every subsequent launch. He would watch the launch and send a beautiful one-page letter with compliments. I always made it a point to share the contents of the letter with all ISRO employees.

In the new year, we presented a challenge to the Master Control Facility (MCF). Till then it had been managing our nine geostationary satellites, and now it was time to welcome the Indian Regional Navigation Satellite System (IRNSS), a series of navigation satellites meant to provide location and timing services within India and 1500 kilometres around the international border. The baseline constellation consisted of seven satellites of which three were to be placed on the geostationary orbit, and the remaining four satellites were to be placed on inclined geosynchronous orbits. The first navigation satellite, IRNSS-1A, was getting ready to be launched and MCF had to manage, within a span of six months, three more new satellites—the INSAT-3D meteorological satellite and GSAT-7 communication satellites scheduled to be launched from Kourou besides the GSAT-14 communication satellite to be lofted by GSLV-D5. Each of these new satellites had to go through a set of prelaunch rehearsals for a fortnight and post-launch operations for three weeks, conducted by a 100-member team of satellite experts. It was a classic case of operations management with a potential resource crunch in the internal resources at MCF and global communication networks. Also, the teams had to be geared up for the mission contingencies of each satellite. That was addressed squarely by a high-level team of satellite experts. D. Ravindranath, who had just been elevated as director of the

Master Control Facility, and his team improvised on the second mission control room. I created a Spacecraft Authorization Board, chaired by Kiran Kumar, with the soul responsibility to oversee and guide the satellite mission operation from its launch to end-of-life, bringing synergy between all players concerned. This was conceptually a parallel to the Launch Authorization Board that we had on the launch vehicle side.

The launch of the IRNSS-1A by PSLV-C22 went off successfully on 1 July 2013. The deputy chairman of the Rajya Sabha, Prof. P.J. Kurien, graced us with his presence at Sriharikota. That day was special for Mini and me, as we had completed three decades together.

IRNSS-1A reached its final abode in the geostationary orbit and its performance met all expectations. India's journey to become self-reliant in satellite navigation had begun. It ultimately owned a regional navigation satellite system that gave accurate position information to users in India as well as the region extending up to 1500 kilometres from its boundary.

But soon we faced difficulties. An event confirmed the fact that there was no room for complacency in the realm of outer space. We were at the Mission Control Centre in Hassan during the launch of the INSAT-3D meteorological satellite on 26 July 2013. In the viewing gallery of the Mission Control Room of MCF, I was accompanied by Dr Alex, Koteswara Rao and some other colleagues. In the control room on the consoles, A.S. Kiran Kumar (chairman of the Spacecraft Authorization Board) and Anjaneyalu (programme director, GEOSAT) were seated with the satellite team at their respective mission consoles. The Mission Control Room housed a number of consoles dedicated to each subsystem of the satellite (e.g. power, propulsion, telemetry and telecommand, control, etc.) which provided the nominal prediction figures as well as the real-time figures from the satellite for that particular subsystem. A display panel synthesized all the key information from the subsystems required for the operation by the mission director and higher-ups.

On that eventful day, we got a confirmation from the telemetry data that the Ariane-5 had perfectly injected INSAT-3D into the intended orbit. We were expecting the confirmation for the solar panel deployment, which was to take place through a time-tagged command. But before we got the confirmation, we lost the telemetry. There was absolutely no communication with the satellite—neither the telemetry nor the carrier signal.

There were a number of activities to be performed on the satellite; the confirmation of which was supposed to be carried by the telemetry. But it was as if the satellite had turned mum.

To deal with such emergency situations, the mission teams had contingency recovery plans, prepared and thoroughly reviewed well in advance. We executed the relevant plan and restored the telemetry. We later learnt that the solar panel deployment did take place as planned, but the satellite was spinning at an alarming rate on all three axes. The gyros, used for measuring the attitude of the satellite, were already saturated and the satellite was slowly going out of hand. Many young engineers were seen with shoulders dropped in despair and chins down with disappointment, apprehending the eventuality.

But the seniors did not lose hope. They tried to break the spin on all three axes using control thrusters; it did not work. They tried to figure out the rate of the spin through radiation patterns of the telemetry antenna. It appeared that the spin rate of the yaw axis was alarmingly high. After ascertaining that, the mission engineers attempted to arrest the rate of the spin through manually operating the control system. The only silver lining was that the solar panels were still facing the sun and hence power availability was not a concern. This was the best available method to rescue the satellite. A few attempts later, the plan started working. After a battle of close to two hours, INSAT-3D was brought under control.

Later we found out that one of the two telecommand receivers and one of the three momentum wheels (used for attitude correction and measurement) had become non-functional.

INSAT-3D was brought into operation after the essential early-orbit operations and testing which went without any problem.

For us this was a warning sign. GSAT-7 and GSAT-14 satellites, which had the same platform (I-2000) as the INSAT-3D, were waiting at Kourou and Sriharikota respectively for launch within a month. These satellites shared similar designs for solar panel deployment and power systems, the triggering points for trouble in INSAT-3D. It was imperative to locate the root cause of the anomaly in INSAT-3D and make corrections in both these satellites. An expert team led by Dr Alex and Shivakumar did it marvellously.

Time of Honour

Amidst this hectic schedule, we were quietly approaching August 2013. And then Arianespace, our European launch service provider proposed to schedule the launch of the GSAT-10 satellite on 29 August 2013, my sixty-fifth birthday. The previous day, my tenure as chairman was to come to an end. My close associate Prahlad Rao, the then director, Satellite Communications Programme at the headquarters, came to my cabin with his characteristic broad smile and inquired, 'Shall I ask Arianespace to prepone the launch by a few days?' I said, 'Not necessary. I am sure ISRO will have a chief on 29 August.'

There was a reason behind my unhesitant reply. By mid-January 2013 itself, I had formally requested the PMO to initiate the process to identify and groom my successor for a seamless transition at ISRO from 29 August 2013. As a follow-up of the succession plan submitted two years earlier, two centre directors, Kiran Kumar and M.Y.S. Prasad, had been elevated by the government to the grade of Secretary to the Government of India before they crossed sixty years of age in 2002 and 2003. They had handled decisive roles—Kiran Kumar had been the chair of the Spacecraft Authorization Board and vice chair of the ISRO council and Prasad had been the chair of the Launch Authorization Board.

As a motion build-up exercise, a couple of special presentation sessions were arranged at Delhi to ensure the visibility of all likely candidates; I deputed them by turn to represent ISRO at key meetings in Delhi. But the government decided to extend my tenure by a year (i.e till 28 August 2014) citing an amendment of law (2007) that provided extension of retirement age to secretaries in departments of space and atomic energy till the age of sixty-six years. This amendment was done to ensure effective implementation of vital national programmes that had long gestation periods. The communication for the extension of my service came in the first week of August 2013 itself, while I was at Sriharikota preparing for a critical attempt at a GSLV-D5/cryogenics flight test.

After toiling for more than four years, the GSLV-D5 rocket, with a redesigned and modified Indian cryogenic stage, was getting ready to launch the GSAT-14 communication satellite. However, at the penultimate stage of the campaign, on 19 August 2013, we had to abort the launch and postpone the mission operation. The launch of the GSAT-7 communication satellite (on board Ariane-5) from French Guiana on 29 August 2013 went ahead in due course and so did the post-launch operation of GSAT-7.

The service extension by a year gave me the opportunity to lead ISRO steadfastly for landmark events such as flying GSLV-D5 with our own cryogenic stage; the launch of the Mars Orbiter Spacecraft by PSLV-C25 in November 2013 and its crucial trans-Mars injection within a month. I felt I had to repay the nation's decision makers for the faith they had shown in me by completing these programmes, as they had very high national significance.

25 January 2014 (Saturday) became a red-letter day in my life. I was just through with my morning prayers, when the telephone rang. It was the joint secretary of the home affairs ministry, Satpal Chouhan. He had called to convey the government's decision to confer on me the Padma Bhushan. By the evening, it was confirmed that besides me, three more had been selected for other Padma awards from ISRO. A.S. Kiran

Kumar, M. Chandradathan and M.Y.S. Prasad had been awarded the Padma Shri. Congratulatory messages poured in. This was a unique recognition for the entire ISRO team and I dedicated the award to all of them. There was jubilation at my alma maters. A few days later, I was invited to participate in the morning assembly at the National High School, where I had spent ten years as a student. I was overcome with nostalgia for my days at the school, and I paid tribute to my beloved teachers. I became the third Padma Bhushan awardee from our vicinity, following the celebrated master of the classical Sanskrit play *Koodiyattam*, Ammannur Madhava Chakyar (2003) and the founder-principal of Christ College, Rev. Father Dr Gabriel, CMI (2006).

Mini and Nilanjan joined me to receive the third highest civilian honour by the Government of India, at the Central Hall of Rashtrapati Bhavan, from the President of India on 26 April 2014.

In April 2014, I was bestowed with an international award, Allan D. Emil, for my outstanding contribution to the field of astronautics, from the International Astronautical Federation, the most revered professional body for space professionals across the world. In the last week of September 2014, I landed in Toronto to receive the award (just after the Mars Orbiter Spacecraft successfully entered the targeted orbit around Mars). The award ceremony took place in the opening session of the International Astronautical Congress. I was seated in the front row of the beautiful auditorium, waiting for the event to begin, when I saw my predecessor, Mr Madhavan Nair, make an entry in his capacity as president of the International Astronautics Academy. I stood up, walked to him and greeted him by shaking his hand. In a rare display of affection, he gave me a close hug, rekindling a camaraderie of forty years.

In many ways, 2014 was a great year for ISRO. The erratic GSLV was tamed by a fine performance by our Indian cryogenic engine. We celebrated New Year's with the success of the GSLV-D5. The Mars Orbiter Mission was safely on its way towards the red planet. ISRO was in excellent shape. I felt the time was right, and I approached Prime Minister Dr Singh in the last week of March 2014 with a request to initiate the process for the transition of leadership in ISRO from 29 August 2014. Within a month came the government's decision to extend my service up to December 2014 on functional grounds and in the public interest. The seminal event of the Mars Orbiter Mission to capture the Martian orbit was slated for 24 September 2014, and probably wisdom weighed towards avoiding a transition at ISRO just three weeks ahead of it. I was happy as I had the opportunity, again, to see to the logical conclusion of a historic initiative taken in my first year as the chief of ISRO.

With ISRO surging strongly ahead, it was time for creating some international alliances for newer capabilities. Soon after the launch of the IRNSS-1B by PSLV-C24 on 4 April 2014 and its safe placing in the final orbit, I led a high-level delegation, comprising Kiran Kumar, Shivakumar, and Sudarsan Srinivasan, to the US and Europe to forge future relationships with NASA, Goddard Space Flight Center, Jet Propulsion Laboratory and Caltech, besides exploring international alliances in satellite communications. At the Graduate Aerospace Laboratories of the California Institute of Technology (GALCIT), we were taken to the cabin of its distinguished alumnus, Prof. Satish Dhawan, which had been neatly preserved for posterity. It was a moment of pride for both ISRO and Caltech to institute a Satish Dhawan endowed fellowship at GALCIT.

Change of Hands

At home, transition in the Union government was taking place in May 2014. Prime Minister Dr Manmohan Singh had held the post of minister-in-charge of space for ten years, during which fifty-seven of India's 113 space missions (till then since Aryabhata of 1975) were executed. India's first lunar mission, Chandrayaan-1, was executed during this phase. Also, the Mars Orbiter Mission had been announced by him in 2012 and we were well on our way to accomplishing the historic feat within the next four months. As a matter of courtesy and my individual respect, just three days before he was to resign from his post, I visited him at '7RCR' to express ISRO's gratitude for his sterling guidance. I felt he had been a source of constant encouragement during those high performing years of ISRO. For me, Dr Manmohan Singh has been a guru, only next to Prof. Satish Dhawan.

As luck would have it, his successor Prime Minister Narendra Modi had always been an ardent promoter of space applications for governance and development ever since he took charge as chief minister of Gujarat in October 2001. Interestingly, in March 2011, when Dr Manmohan Singh had visited SAC, Ahmedabad, we had witnessed a unique scene—Chief Minister Narendra Modi narrating his first-hand experience on using communication satellite for education and remote sensing data for land management in Gujarat to him. That day he acted as one of the ambassadors of the Indian space programme. I have often given Gujarat's example as the role model for space applications in India.

The new prime minister granted me an audience on 29 May 2014, on his third day in office, at the corner room of the South Block. I was overwhelmed by his warmth and personal touch as he greeted me inside his chamber. I presented him with a signed copy of Prof. U.R. Rao's book *India's Rise as a Space Power*. I was well prepared for this maiden meeting, armed with a concise booklet that I had prepared on 'India in Space-Extending Limits and Exploring Possibilities' along with a few fliers on the imminent space missions and a model of the Mars Orbiter Spacecraft slated to be placed in the Martian orbit within the next four months. The interaction, which lasted fifteen minutes, was quite satisfying. He listened carefully and also asked sharp questions in the end. What more could ISRO have expected? I also met the new MOS (space and PMO), Dr Jitendra Singh, the PM's principal secretary, Nripendra Misra, and later the national security adviser,

Ajit Doval, in this regard. The new team at the PMO offered their best support to ISRO.

At Sriharikota, the PSLV-C23 was just one month away from launching SPOT-7, another French remote sensing satellite, and four co-passengers from Canada, Germany and Singapore. The PM graciously accepted our invitation to witness this launch from Sriharikota, scheduled for 30 June 2014. The entire ISRO family was excited about the presence of the new PM. Sriharokota was bubbling. Soon after his arrival the previous evening, the PM had taken a detailed review of the activities of the Department of Space and had interacted with the functionaries and ISRO's centre directors present there. It was a lively session with sharp questions and excellent responses. Later that evening, the PM reached the launch pad accompanied by the Union minister, Venkaiah Naidu, and other dignitaries; our PSLV-C23 was blazing with floodlights around. M.Y.S. Prasad and the mission director, Kunhikrishnan, briefed them on the launch preparations. This was followed by a thirty-minute walk-through of the nearby Vehicle Assembly Building where the PM mingled with engineers from the launch vehicle domain including K. Sivan and S. Somanath, and took time to shake hands with the technicians present there. At the end of the visit, I made my customary briefing of eventualities in a launch, requested him to permit me to be at the hot seat at the mission console during the launch and proposed Dr Suresh to brief him at the VVIP gallery throughout the proceedings. As before, all my demands were happily accepted by the PM.

The next morning, we had a successful launch. Prof. U.R. Rao and Dr Rangan were present at the VVIP gallery and, as planned, Dr Suresh briefed the PM on the progress of events and the performance of the launcher during the flight. After the launch, he made it a point to greet everyone present at the mission consoles.

The PM's invigorating speech from the Mission Control Centre demonstrated his depth of understanding of the potential of space technology for governance, development and diplomacy. Needless to say, ISRO's plate was full with activity!

By the time I returned to Bengaluru via Tirupati, the pleasant order reconstituting the Space Commission had arrived. Kiran Kumar was inducted as a member of the Space Commission as per a recommendation that I had made a fortnight ago.

I started preparing for a happy landing within six months. I prepared a to-do list that I wanted to complete in the next six months before leaving the pedestal. It read as follows:

(i)   Orbit Insertion Operation of the Mars Orbiter Spacecraft
(ii) Launch of India's third navigation satellite, IRNSS-1C, on board the PSLV-C26 launcher
(iii) Launch (from French Guiana) and early orbit operation of GSAT-16 communication satellite
(iv) Conduct an experimental flight of India's advanced launcher, LVM3, to gauge its performance in crucial atmospheric phase, and use this opportunity to test recovery of an Indian crew module.

And of course, I decided to have a graceful exit, not to meddle with the organizational priorities under my successor, and more importantly not to take a decision that would tie the hands of my successor. I was getting mentally and institutionally ready for the exit.

24

A PERFORMING ARTISTE

The evening of 15 December 2014 was an unforgettable one at the Madras Music Academy, where I had the rare privilege of inaugurating the eighty-eighth annual conference, celebrating some of the most distinguished artistes in the Carnatic school of music. The annual conference of the academy had been the dream destination of every Carnatic musician for decades. It provided the opportunity to listen, to learn, to emulate, to perform and to be recognized in the circuit. On the dais, Sangita Kalanidhi awardee Mr T.V. Gopalakrishnan was flanked by two Sangita Kalanidhis—Mr Umayalpuram K. Sivaraman, the living legend adored for his mastery on the mridangam and Ms Sudha Raghunathan, the versatile and celebrated Carnatic vocalist gifted with a spellbinding voice. A dazzling array of Sangita Kalanidhi awardees of yesteryear, including the legendary violinist Mr T.N. Krishnan, adorned the front rows of the audience. The main hall of the TTK auditorium was flooded with eminent musicians, erudite connoisseurs, young performers and passionate students. I was ecstatic to be escorted by its president Mr N. Murali to the dais. Being a lifelong student and admirer of this school of music, I naturally had goosebumps while addressing the learned audience. My inaugural address, delivered partly impromptu, was received with nods of approval and applause from both sides of the dais. I concluded by underlining the need for protecting, preserving and promoting the classical form for posterity. I was happy when a music reviewer cited, a fortnight later, my suggestions on the tempo.

We were returning from Sriharikota after the prelaunch review of the first experimental flight of LVM3, accompanying me were two of my close associates who were inclined towards classical music. The project director of PSLV, P. Kunhikrishnan—well known for leading a dozen successful PSLV launches—was an amateur Carnatic flautist. And of course there was Nilanjan, trained formally in tabla during his childhood.

We returned to our guest house in Adyar and later gathered for a small post-dinner chat. Kunhikrishnan commented, 'Sir, the talk was amazing! Did you prepare for it or was it impromptu? The quotes, the content and the message were apt and scholarly.'

I smiled and said, 'Well, I have been at it for the past six decades; people have different ways of relaxing and getting rid of the chaos in their professional lives. Some choose sport, some like to socialize, some like to keep to themselves, I had this passion for learning music. It is of course more than a means of relaxation; it is a form of sadhana (meditation) for me.'

'But you are trained in classical dance too, isn't it?' Nilanjan intervened.

'That's quite a story. You see, I grew up around the holy precincts of the Koodalmanikyam temple in the ancient town of Irinjalakuda in Thrissur, the cultural capital of Kerala. The *koothambalam* (temple auditorium) of the Koodalmanikyam temple comes alive for forty-one days during the months of May–June with Koothu and Koodiyattam performances by the renowned *chakyars* (priestly performers) of the Ammannur family, the most ardent followers of Bharata Muni's *Natya Shastra*, an ancient Indian text on performing arts. They were our neighbours too. The temple shares its wall with the Unnayi Warrier Smaraka Kalanilayam, a Kathakali school set up in 1955. Unnayi Warrier, a traditional garland maker for the deity at Koodalmanikyam temple in the eighteenth century, had authored a unique romantic Kathakali play called *Nalacharitham*, known for its literary richness and musical quality. This had been a touchstone for performing artistes. During the ten-day-long annual festival at the temple, I could be spotted next to percussion artistes, who would be performing *Panchari Melam*, a complex rhythmic pyramid of five tempos from a slow ninety-six-beat cycle to a superfast six-beat cycle. I used to try it out on a wooden plank; I still wonder why I did not pursue percussion!

'It was only natural for me to be drawn into the world of performing arts in my early childhood. At the age of seven, I debuted on stage with a folk dance team during the school's anniversary day celebration. I had the ability to grasp and demonstrate with ease the moves taught by our music instructor, Mrs Easwari, while she was teaching other students. This innate ability gave my parents the confidence to put me for formal training under Tripunithura Vijayabhanu, an exponent of "Kerala Natanam". That was in 1959.'

'What exactly is "Kerala Natanam"?' Nilanjan inquired.

'It is a distinct dance style carved out of Kathakali in the 1930s. A legendary Kathakali dancer, Guru Gopinath, was its innovator. He did away with the intricate make-up and costumes of Kathakali, leaving intact its classicism of body movements, gestures and facial expressions.

'For the next two years, I took dance lessons from Vijayabhanu Sir every weekend, in a long dining hall adjacent to the Koodalmanikyam temple. There were about twenty students in the age group of eight–fifteen years at any point of time. After a few months of training, Vijayabhanu Sir asked me to demonstrate the basic steps to my juniors; it was an opportunity for me to perfect my art. We were taught appealing scenes from Kathakali and mythological stories as well as stories woven around social themes such as lives of farmers, fishermen and hunters. I was first exposed to the *Bhagavad Gita* during the practice session of *Geethopadesam*, where I played Arjuna. I had to portray the plight of a valiant Arjuna falling into a depressive state of mind, only to be rejuvenated by the advice of Krishna. Recognitions and awards flowed in as I matured in this art form, and it occupied a special place in my heart.'

Kunhikrishnan was curious, 'When did you shift to Kathakali?'

I said, 'It was in 1962. That was the need of the hour in our school and I relented. A Kathakali dancer, Raghavan Asan, and the instructors at the Unnayi Warrier Smaraka Kalanilayam, steered me towards the dance form. The hand gestures and facial expressions deployed in Kathakali to convey each phrase had a definite start point, a sequence and an end point, all to be fitted within the metric pattern of the rhythm. My expressive face and adherence to rhythm made this transition smooth and fast. Another doyen of Kathakali dance, Pallippuram Gopalan Nair, the then principal of the Kalanilayam, supervised my training for the portrayal of the scene depicting Damayanthi's interlude with the celestial swan, the emissary of her beloved King Nala. This was one of the prominent sequences in the Kathakali play *Nalacharitham*.

'Unfortunately, rheumatic arthritis had me in its grip by then and I bade goodbye to Kathakali in 1964.'

'A break when you had just started flourishing,' Kunhikrishnan observed.

'Yes, you could say that. It was an excruciating phase for the artiste in me. I could not perform but I wanted to continue the lessons. After doctors allowed me to move around, I became an ardent observer of performances. You have probably heard the term "Zeigarnik effect" which means "people remember incomplete or interrupted tasks better than completed tasks". I guess this was kicking in me. I observed and absorbed the performances of several masters, and tried to evoke them in my performance sixteen years later, when I danced again on a stage.

'I performed *Shiva Thandavam* in 1979 at a Durga puja pandal organized by our Bengali friends of VSSC in Thiruvananthapuram. After I moved to Bengaluru in 1981, I came across T.V.A. Varier of Hindustan Aeronautics Limited or HAL, a professionally trained Kathakali dancer. I came under his tutelage. Among many others, he taught me the role of Parasurama, the ferocious character who finally gets placated after an encounter with Rama. I essayed the same role at the Koodalmanikyam temple festival in 1982. That was a kind of rebirth of the dancer in me. I was a part of many renditions after that. In mid-1980, I played the role of Bheema in *Kalyanasougandhikam*, Daksha in *Dakshayagam* and Hanuman in *Lavanasuravadham*. In the 1990s, I switched over to roles that did not require elaborate costumes. My last Kathakali performance was in December 1995 at Jalahalli Sree Ayyappan temple (in Bengaluru) portraying a Brahmin in *Santhanagopalam*. Now that I am at the fag end of my career, I want to go back to the stage to perform some of these roles again.' Nostalgia engulfed me as I relived my journey as an artiste.

'That's a terrific plan for your post-retirement life!' Nilanjan supported my wish and came up with a significant observation, 'But when did you start training in vocal music? When you were in IIMB?'

I said, 'You are right. I was twenty-six years old then, a bit old to start learning a new art form. To be frank, I didn't have a very melodious voice when I was young. When I look back, I realize that it was monotonic. I could not pick up music even though there was a lot of music around in our school with seniors like P. Jayachandran, who later became a renowned playback singer in Kerala. I guess the urge to express took the form of singing in my mid-twenties, as dance was not within my reach by then. I learnt and kept learning more, and I think I have reached some level now. I was invited to perform for ninety minutes at the Bengaluru Sangeet Utsav just a fortnight back. It was an arduous path, perseverance brought me this far.'

Kunhikrishnan remembered, 'Yes I saw a few pieces of it on the *India Today* TV channel. It was fantastic.'

I smiled at his appreciation and said, 'Somewhere in 1975, I could feel signs of modulation in my voice and attempted simple songs. I was a student at IIMB then. Vidwan H.S. Radhakrishna, a violinist, agreed to coach me, and I was introduced to the basics. I struggled with the help of a harmonium to stretch the range of my voice and crossed one octave. When I went back to Thiruvananthapuram, the lessons continued with Vidwan N.K. Kolappan Pillai for a couple of years. However, it did not yield much result for want of practice from my side. I was staying in a lodge in those days and that made it difficult.

'In June 1978, I moved to an independent house in Sreekaryam, and that's when I started learning music seriously. Prof. Vechoor Harihara Subramania Iyer, a leading musician and a disciple of the doyen Semmangudi Srinivasa Iyer, accepted me as a disciple. He was a great teacher. Even though I had official engagements at VSSC, I showed up religiously at Vechoor Sir's house at Valiyasala near the railway station at 8 a.m., three days a week, for the ninety-minute session. He taught me several new compositions. In fact, his lessons are imprinted in my memory. We continued this routine till 1981. I was treated like a member of his family. I still have the music notes scribed by him for those sessions. I first sang on stage during a festival at the Koodalmanikyam temple in April 1980 where I rendered my favourite composition, *Viriboni,* in the raga *Bhairavi*. My performance, which lasted fifteen minutes, was an opening act for Vechoor Sir's concert.

When I moved to Bengaluru in 1981, I sought the tutelage of Vidwan R.K. Srikantan (a Padma Bhushan recipient in 2011). I approached him with an introductory letter from Vechoor Sir and he readily accepted me as his disciple. He was a demanding teacher, and would move on to a new lesson only after making sure that the disciple had picked up the previous ones without any slip. He had a unique style of presenting his music (*bani* and *patanthara*) which left a deep impact on his disciples. His son and my friend, Vidwan R.S. Ramakanth, tutored me at times and taught me several beautiful compositions. This process continued till I moved to Hyderabad in 1997. I had the rare opportunity of sitting by his side during his assiduous practice sessions, assisting him during concerts, and giving vocal support on a few occasions. I'll cherish these memories forever.'

Nilanjan added, 'I was amazed to meet him at his residence in 2012 when Chairman Sir was taking music lessons. He sang for almost an hour at Antariksh Bhavan during the Kannada Rajyotsava celebrations in December 2013. His voice texture and stability was beyond imagination, even though he was more than ninety years old then. Sir, it was great to see you sing along with him and his son.'

I said, 'That was just forty days before he breathed his last. I also had the fortune of knowing another legend, Dr Nookala Chinna Satyanarayana (Padma Bhushan recipient in 2010), during my stint in Hyderabad. We met in June 2001. Dr Nookala was a great musician with razor-sharp analytical skills. His style of singing and presentation of compositions were quite distinct. Nookala Sir introduced me to nearly forty compositions, mostly complex ones, by Saint Thyagaraja. Practising and mastering them is still an agenda.'

Kunhikrishnan was eager to know more. 'Sir, we all know how demanding your life was at Bengaluru; how did you balance your music and official engagements?'

I said, 'It is quite simple. I used to earmark an hour in a day, on average, for my music practice. In the car I would listen to Carnatic music and would often sing along too. Music is one of the biggest stress busters. See, at times you need to just switch on and then switch it off. It has come to my rescue during trying times. I have often turned to music to regain my composure. These music sessions have also strengthened my memory.'

Nilanjan asked, 'Did your pursuit of music influence your character and inner strength?'

I expressed with conviction, 'Yes, it did. When we begin, most performing art forms are means of entertainment and self-expression. Slowly, they evolve as a means of self-realization and later bring in the realization of the super power.

'Indian music revolves around the triad of aesthetics, grammar and spirituality. The spiritual aspect of music gives ultimate bliss to the musician and the listener. This is not only a means of relaxation; this is something divine, probably a way to reach out to the super power. Let me give you an example. I had taken a commitment to participate in the annual festival at Koodalmanikyam temple and render classical music for an hour there on 26 April 2010. I was in a dilemma as just a fortnight earlier, the first flight during my tenure as chairman, GSLV-D3, had failed. Finally, I flew down to Cochin for a day and honoured my commitment. That evening, I sang *Makelara Vicharamu*, a composition by Saint Thyagaraja, in the raga *Ravichandrika*. In the composition, a devotee asks lord Rama why he should worry when the world is a stage and the strings of destiny are controlled by the lord himself. In fact, a video of my performance was uploaded on YouTube by someone in the audience.

'Over the last two decades, I have had plentiful opportunities to sing, mostly at temples. There were times when I felt as if I was being pulled to another realm. This is something one has to experience and cannot be expressed through words. These experiences followed a pattern. In these instances, I invoked my guru and god through my soulful singing with the focus on the meaning and context. Probably that was my way of meditating. It is a state when one doesn't feel anything other than the divine presence.'

'Wow! I have heard people talk about being in the zone but this is probably as close as you get,' Nilanjan said. 'Thankfully the media has stopped joking about your temple visits these days,' he added.

I smiled. 'See, it is their job to ask questions, relevant or not. I am often asked about my state of mind as I sit at the mission console for a launch, with the entire nation watching. Many lessons from the *Bhagavad Gita* tell you to do the duty assigned, without worrying about happiness and distress or loss and gain. Indeed, we do our best. But the results are not in our control. That is the time when faith helps you in maintaining your composure and taking the right decision. This is, of course, my faith and opinion and, like anybody else, I am entitled to it. That is why after giving the go-ahead to the mission team for every launch, I go in search of my roots to re-establish my faith and pray at several temples. I do this at my personal expense. This is the right and freedom assigned to me like any other citizen of India. I guess, of late, the media has understood this. You know they interviewed me on a very positive note when I was on my fifty-second visit to the Sabarimala shrine. This was just after the launch of the Mars Spacecraft.'

'You are absolutely, right,' my audience concurred.

The clock struck 12 a.m., and we knew we had to sleep to catch the early morning flight. A rigorous day at the office was awaiting us.

25

TAMING THE NAUGHTY GSLV

Each time a launcher threw a tantrum, ISRO held congregations of its engineers to find ways and means to counter it. There had been several instances of ISRO successfully scripting the diagnosis and coming up with solutions. It happened when SLV-3 stumbled in 1979 and ISRO successfully launched it a year later. The practice continued when ASLV failed in succession in 1987 and 1988. These 'successful failures' taught us fundamental lessons in rocket technology, and ISRO flew ASLV successfully in 1992. History repeated itself when PSLV failed in its debut flight in 1993; it was the learning of a lifetime for PSLV architects. The subsequent successful flights of PSLV since 1994 stood testimony to the efficacy of this collective, open-minded, and diligent analytical process by the engineers of ISRO in the past three decades.

On an overcast afternoon of 14 June 2011, the foyer of the main building of the Vikram Sarabhai Space Centre was abuzz with 500-odd engineers from VSSC, LPSC and ISRO Inertial Systems Unit and about twenty-five erudite aerospace experts and academicians attending the technical review on GSLV with the Indian cryogenic stage. That was a congregation of several generations of rocket engineers with the common agenda: 'to successfully fly the GSLV, powered by the Indian cryogenic upper stage'.

Mining the Details of Previous Failures

The chequered history of GSLV had baffled generations of rocket engineers at ISRO. We sat down for a rigorous and transparent introspection of the past seven flights of this launcher. Its underperformance during its debut in 2001 had been assessed and that had led to its success in 2003 and 2004. Its fortunes reversed in 2006 when a gas generator in one of the L40 strap-on stages ceased to perform and that resulted in a spectacular failure. An error of one millimetre while machining the control port of a flow regulator slipped attention at every step of inspection and testing. The recovery flight of 2007 could, at best, be termed an erratic one. GSLV had also suffered from a malfunction of the control system of one of the strap-on stages (L40). We suspected that the culprit was a gas motor that was driving the control actuator, though this failure hypothesis could not be convincingly proved.

Three years later, in April 2010, we conducted the much-awaited flight test of the Indian cryogenic upper stage. In this flight, all the L40 strap-on stages, the S139 solid core stage and the liquid second stage performed excellently; then came the turn of the cryogenic engine, which ignited. But the fuel booster turbopump of the cryogenic stage stopped a second later and the mission failed. The two plausible failures were hypothesized—seizure of rotor and rupture of the turbine casing—both could not be convincingly demonstrated in the simulations using similar hardware; yet essential design changes were made to avoid their recurrence. The next failure of the GSLV in December 2010 (GSLV-F06) was a brutal manifestation of the inherent flaws in the launch vehicle. We realized that in all the previous GSLV missions, a suction pressure existed at the location of the shroud, fixed to the bottom portion of the cryogenic stage, inducing structural loads at that location. This flaw did show up in September 2007 (GSLV-F04) when one of the connectors got detached, but it was dismissed as a minor deviation as the connector was not carrying any control signal. This manifested heavily in 2010 when one of the connectors carrying the control signal got disconnected and we lost the mission.

It was obvious that we did not have the luxury of another failure involving a GSLV or any other mission! It was, in a sense, the lifeline mission for ISRO; the two decades of work, initiated by Prof. U.R. Rao, was about to bear fruit.

During the review of June 2011, we had among us two pioneering centre directors of VSSC and LPSC—Dr S.C. Gupta and Dr A.E. Muthunayagam— who were at the helm during those decisive early years of the 1990s. Both heavily contributed to the discussion and gave useful inputs.

We all had several questions lined up and we tried our best to find out answers to the following:

(a) What are the remnant issues and hidden failure modes in GSLV?

(b) Have we understood the root causes for the failure of the Indian cryogenic stage?

(c) Had the fuel booster turbopump worked, would some other flaw have shown up during the rest of the flight?

(d) What were the drawbacks in the quality control and reliability management set-up?

We literally dug out all the post-flight and failure analysis reports of all previous GSLV flights to find the flaws that led to the non-fulfilment of the mission.

Over the past year, the complexion of the GSLV and cryogenic project teams had changed. It started with S. Ramakrishnan moving out of VSSC and taking over the post of centre director of LPSC in June 2010. LPSC had held a pivotal role in the buildout of the GSLV; it was the main centre for development of the Indian cryogenic stage and delivery of the L40 strap-on stages and the second stage, GS2. He was the best in ISRO for that job. The project director of GSLV, G. Ravindranath, had gone through an agonizing stint of seven years. His earnest request for a change had to be honoured when a suitable slot of director of the Inertial Systems Unit opened up in April 2011.

We needed an apt successor for Ravi who would understand the tricky GSLV launcher in its entirety and also drive it in fast-track mode. One of the senior engineers from VSSC came forward and volunteered to take up the challenge. I was considering a few others from both VSSC and LPSC as well. And then my gut instinct told me that K. Sivan, the then deputy director of the aeronautics entity of VSSC and the project director of the Reusable Launch Vehicle-Technology Demonstrator, could be a good choice. Sivan, by then, had gained rich experience in control and guidance as well as in mission design, simulation and synthesis of both PSLV and GSLV. Most importantly, he had played a key role in the failure analysis of the GSLV-F06 during the past three months. I consulted Dr Suresh and he endorsed my proposal at once. I also consulted the director of VSSC, P.S. Veeraraghavan, who also agreed. Finally, I requested Veeraraghavan to break the news to Sivan and persuade him to take up the task. Sivan happily took up the challenge. We also armed him with two ebullient associate project directors—Umamaheswaran and M. Mohan.

But more changes were in the offing. A few months later, Sasidharan Nair, the project director of the Cryogenic Upper Stage Project was retiring. We identified N.R. Vishnu Kartha, a warhorse from LPSC, to take over. The symbiotic relationship that Sivan and Vishnu Kartha forged surpassed my expectations at the time of catalysing their synergy for GSLV. V. Narayanan was promoted to the post of associate vehicle director. This core team of GSLV was an amazing amalgamation with a great sense of mutual respect and they depended on each other's strengths. Their objectivity brought out the best from all concerned at all levels of ISRO during the next three years.

Coming back to the review of June 2011, the project director, Dr Sivan made a systemic assessment of the GSLV launcher—a synthesis of the lessons learnt from past failures and a possible configuration of the GSLV-D5 that could be flown next. To bring out all the hidden but possible flaws, we organized a post-dinner brainstorming session of 'young turks' anchored by R. Hutton

and Sowmyanarayanan, two live wires from VSSC. The experts
and veterans gave their analytical interventions.

The next task was to freeze the configuration of GSLV-D5
and the GSAT-14 satellite to be flown with it. Some of the experts'
suggestions were technically appreciable but involved an element
of risk that we could not afford at that juncture. They had to
be moderated with ruthless managerial prudence and I did this
politely. One such crucial decision was sticking to a 3.4-metre
diameter for the heat shield of GSLV, which had been raised to
4 metres during GSLV-D3 and F-06. Although, it was not proved
if the larger heat shield had any bearing on the failures, we decided
to go back to the previous dimensions. We knew for sure that a heat
shield with a 3.4-metre diameter was sufficient to accommodate
our satellites—up to 2200 kilograms. Hence, there was no point in
putting unnecessary strain on the GSLV from a larger heat shield.
Next, we limited the GSAT-14 satellite mass to 2000 kilograms.
This was a conscious conservative strategy to ensure its injection
into the specified orbit even if the cryogenic stage failed to perform
to its full capacity. If the mission was successful, the GSAT-14 could
add twelve communication transponders to national capacity;
testing new technology elements such as fibre optic gyro, active
pixel sun sensor, Ka-band beacon were the icing on the cake. The
cardinal goal was to put the satellite into the right orbit; stretching
its limits was relevant only when this first step was complete.

I was convinced that ISRO could go ahead with the GSLV-D5
after the essential tests and reconfirmations were completed on
the ground.

Afterwards, it was truly a sprint of thirty months towards
a single goal—fly GSLV successfully with our own cryo. That
was when the tenacity of the team came to work, to scale a new
summit of the Indian space programme.

We reviewed and revisited the design of the GSLV launcher,
its aeronautical and aero-thermal design, the mission sequence,
trajectory design, estimation of aerodynamic and acoustic loads,
electronics and control systems, software formulation, fluid

handling systems and procedures of checking the assembled launcher. Besides, we also redesigned the connector mounting scheme, shroud and wire tunnel that had failed us in December 2010. Computational fluid dynamics analysis was a handy tool which Ashok and the team at VSSC had mastered by then. To complement it through the experimental route, the aeronautics team of VSSC did around 950 test runs in wind tunnel tests (both in India and Russia) with scaled-down models of GSLV before and after design modifications. Sivan's vast experience helped him to guide these tasks from the beginning.

The fuel booster turbopump was the suspect in April 2010. Besides the design changes to avert the two plausible failure modes of seizure of rotor and rupture of turbine casing, we needed to devise a test matrix after its assembly to ensure smooth functioning in the low temperature of the cryo environment. We established a new test facility at Mahendragiri by June 2011 to create that environment.

During the flight, ignition of the cryogenic engine and two Vernier engines takes place in hard vacuum conditions encountered at 130 kilometres altitude. These engines have to operate for nearly twelve minutes at the last stage of the flight. Our existing test facility at Mahendragiri could test these engines but could not simulate the hard vacuum conditions around. We were then in the final leg of building a High Altitude Test Facility (HAT) for the CE-20 engine (meant for the bigger LVM3). In February 2012, we decided to customize it for testing the cryogenic engines of GSLV. This task was taken up on top-priority basis and the high-altitude simulation tests were done in March 2013. The tests were successful and they validated the theoretical models of the cryogenic engine and stage and the timing sequence leading to successful ignition and its sustained firing.

Mathematical modelling of the cryogenic engine and stage had to be done meticulously, almost from scratch. V. Narayanan's team instilled tremendous confidence in us as their model was successfully used to explain the failures that occurred earlier.

Our overall initiative of zero defect delivery systems from shop floors to the Mission Control Centre proved crucial for GSLV.

Another failure mode in the GSLV-D3 was identified through the 'fishbone analysis' carried out by a team led by M.S. Suresh. This theory hypothesized the possibility of contaminants in the fuel line that had stopped the functioning of the fuel booster turbopump. We could not prove the theory in the laboratory at the time, but we wanted to ensure that this probable reason was also taken care of. We looked for all possible sources of contamination, starting from the assembly process at Mahendragiri, transportation containers and the storage systems at both Mahendragiri and Sriharikota.

Finally, we had our eyes on a component called the propellant acquisition system. This was imported and stored in a sealed container. It struck me to conduct a vibration test on this stock component. Eyebrows were raised, but Sivan latched on to it. To cut the story short, we found foreign particles as contaminants. We could confidently say that the root cause for the stoppage of the fuel booster turbopump was the contaminant. In a way, we felt elated and confident to go ahead. Our mechanical engineers quickly designed a new propellant acquisition device and fabricated it at jet speed for use in the flight stage.

A High Level Review Committee chaired by me was put in place to oversee the 'thirty-month sprint'. We met unfailingly once a month and more frequently when needed for close monitoring and timely managerial interventions. Sivan and Vishnu Kartha convened them and the minutes of the meeting were jointly cleared by directors of VSSC and LPSC. It became a self-sustaining and self-correcting system.

While the activities were at their peak, I made it a point to communicate with the entire organization frequently. I felt that when the teams were being pushed to the limits, they had the right to know the bigger picture. I often visited the launch vehicle centres/units, and various labs, and communicated with the engineers—lending them a patient ear and of course taking back their suggestions for further improvements.

Prime Minister Dr Manmohan Singh had been keenly following the progress of the GSLV and he had a question or two on it every time I met him. I upheld my philosophy of transparency and accountability as I kept the Space Commission and the concerned officials in the government in the loop, besides keeping the media updated about the complexities, problems and our action plan. I was tackling an onerous task on behalf of the country, and I was fully aware of the public's opinion on the GSLV and the need to ensure visibility and accountability associated with it.

Concurrently, the Flight Readiness Review (chaired by M. Annamalai) and the Mission Readiness Review (chaired by Dr Suresh) started the proceedings for the GSLV-D5 flight. Each problem was discussed in detail until acceptable explanations for the observations were provided by the designers. I recall several animated discussions around induction of polyimide pipelines of complex geometry developed by the polymer group of VSSC. A number of additional tests were suggested by these committees and all were carried out without fail. The flip side of those sessions was that I learnt a lot about the latest advancements in rocketry.

The seniors too showed interest. Prof. U.R. Rao and Dr Rangan were gracious to come down to Thiruvananthapuram during June–July 2012 to listen to the teams on the status of corrective actions and tests. Dr Gupta, Dr Muthunayagam, Mr R. Aravamudan, Dr P.S. Goel, Dr T.K. Alex, Dr S. Vasantha, Mr N. Vedachalam, Mr R.V. Perumal, Mr K. Narayana and many other veterans of ISRO stood with us through this challenging period.

Finally, we started the assembly of the GSLV-D5 on 31 January 2013 at the Vehicle Assembly Building near the second launch pad at Sriharikota. The flight model of the cryogenic stage, assembled at Mahendragiri, was moved to Sriharikota soon after the successful testing of the cryogenic engine in simulated high-altitude conditions in May 2013.

After all these excruciating efforts, ISRO needed to reassure itself and the rest of the country on the preparedness of the

GSLV-D5 launch vehicle with special focus on the Indian cryogenic stage. Once again, I invited the National Panel of Experts for a technical review on 23 July 2013. After a day-long session at Antariksh Bhavan, the panel expressed its confidence in the analyses, design modifications, testing and qualification as well as the quality control mechanisms in place.

In the first week of August 2013, I made a short trip to Delhi and briefed the PM at Parliament House. I assured him that we had done our best. However, we were ready for any eventualities. Finally we zeroed in on 19 August 2013 for the launch of GSLV-D5.

The Penultimate Challenge

The twenty-hour-hour countdown commenced on 18 August and it progressed smoothly. Members of the media were calling me to gauge the mood at Sriharikota. The country, at all levels, was eagerly and anxiously looking forward to this event. People were waiting for the live telecast and those staying close by thronged the Satish Dhawan Space Centre.

Montek Singh Ahluwalia, the deputy chairman of the Planning Commission, was attending the launch. The countdown had reached its final phase. The Mission Control Centre and Launch Control Centre had full attendance. VIPs had taken their seats in the gallery. I had informed the secretary to the President of India of this event, as instructed by him after his visit to Sriharikota. I was informed that he would be watching the live telecast from Rashtrapati Bhavan.

Barely two hours before the scheduled lift-off, just before I was going to leave from the Kalpana guest house for the Mission Control Centre, the phone rang. From the other end, M.Y.S. Prasad, the director of SDSC, said in a grave voice, 'There's a leak, I am giving the phone to Ramakrishnan.' Ramakrishnan had taken over as director of VSSC in January 2013 and Chandradathan had moved as his successor to LPSC. Prasad instantly handed over the phone to Ramakrishnan, who sounded worse, 'The leak is serious.'

'Call off the launch,' I said. Two hours before lift-off, as per the countdown procedure, the propellant tanks of the four strap-on stages and the liquid second stage had been pressurized to their preflight conditions. And then the range operations team had spotted a leak in the second stage's propellant tank on the CCTV camera. Prasad, Ramakrishnan and Chandradathan had promptly instructed their staff to depressurize the tank and had then called me.

As I hung up the phone, the scientific secretary, Koteswara Rao, joined me. He was by my side in letter and spirit. We authorized the mission director, Sivan, to formally call off the launch and all swung into action to safely recover the rocket, satellite and the launch pad. Along with Prasad, Ramakrishnan and Chandradathan, I went to the media room and briefed them on the situation.

At that time, about 750 kilograms of UH 25 fuel had leaked out, leading to contamination of the area around the launch pad. This hydrazine-based chemical had to be pacified before anyone could approach the launch pad even with all safety gears. The GSLV-D5 rocket was loaded with 210 tonnes of liquid and cryogenic propellants as well as nearly 140 tonnes of solid propellant. The leak, as seen in the CCTV footage, appeared to be in the lower portion of the propellant tank or the fluid lines of the second stage.

The outcome of any accidental ignition would have been catastrophic. The entire island of SDSC-SHAR could have been annihilated. But it was deftly handled by the range safety team of SDSC-SHAR and a major calamity was averted without any casualties. I formed a High Level Task Team under Mr K. Narayana (the former director of SDSC-SHAR) to identify the cause of the leak and to work out an action plan for quick restoration of the mission. Two more former directors of the Satish Dhawan Space Centre (Annamalai and Chandradathan) as well as the current director, Prasad, were members of the team. As has been the practice in ISRO, all relevant experts stood by. The concentration of hydrazine emanating from the fuel leak had to be brought down from nearly 5000 ppm to a tolerable level of less than 1 ppm before anyone could even access the launcher to start disarming a host of pyrotechnic devices, disconnecting the electrical connections and fluid circuits connected to the launch tower. This was one of the toughest safety challenges we had to face.

I sat through the proceedings of the task team and inspected several operations with the teams at Sriharikota. The presence of the top management at the site instilled great confidence in the 450-member restoration team of SDSC-SHAR ably led by Prasad. The team worked round-the-clock for six days to move GSLV-D5 safely back to the Vehicle Assembly Building. This was a demonstration of the technical and managerial maturity of the entire ISRO team and their ability to take quick action when the situation demanded it.

Fortunately, the avionics equipment bay and cryogenic upper stage were not affected and they were preserved at Sriharikota in the prescribed ambience. The second stage, GS-2, where the leak had occurred, had to be stripped and taken to Mahendragiri for a detailed inspection and radiography.

The root cause was identified. The leak was due to stress corrosion cracking the propellant tank made of AFNOR 7020 aluminium alloy. In fact that was the last one in a series of propellant tanks made by ISRO in the 1990s for twenty-five flights of PSLV and GSLV. We had followed the prescribed process of material testing, storage conditions and pressure testing before using the tank.

A parallel process had been initiated way back in 2002 to migrate to a superior aluminium alloy, AA 2219, for propellant tanks of PSLV and GSLV. That process had been completed for the propellant tanks of PSLV by 2012 and for the GSLV strap-on stages by 2007 itself. The first propellant tank made out of

AA 2219 for the GSLV second stage had been finalized just a few months earlier. Chandradathan, director of LPSC, quickly got into action to assemble a new second stage for the GSLV-D5 flight.

The propellant tanks and fluid components of the four L40 strap-on stages had got wet from the propellants and hence could not be reused beyond thirty days. However, since their engines had not been exposed to the propellants, they could be reused. Assembling the four L40 strap-on stages within the next two months was an onerous task for our industrial partner, HAL Aerospace Division at Bengaluru. But the chairman, R.K. Tyagi, and his colleagues took up the challenge and ensured timely delivery without any compromise on quality.

The domain experts had differing opinions on the need to replace the S139 solid core stage and the core base shroud of the first stage that housed a plethora of electronic systems and interconnections. We chose to be conservative; managerial prudence and technical diligence prevailed as a matter of extreme precaution.

Four more months of gruelling tasks followed, running parallel with the historical launch of the Mars Orbiter Spacecraft on board PSLV-C25 (in November 2013). Finally, ISRO was ready for the launch of the GSLV-D5 by mid-December 2013. But another challenge came in the guise of nature; three cyclones—Helen and Lehar (19–28 November) and Madi (6–13 December) kept us on our toes all through the final phase.

Moment of Reckoning

On 5 January 2014, GSLV-D5 was ready for the launch at the scheduled time of 1618 hours. GSLV flights had always been stressful events for scientists at Sriharikota and it was obviously heightened a bit more this time. It was truly justified with the previous five bitter experiences, including two failures and one last-minute launch call-off that had occurred five months back.

Not many dignitaries were attending the launch this time, apart from Dr Muthunayagam (a former director of LPSC) and our very own Prof. Yash Pal. We were happy with this setting. After conveying the customary greeting to all the engineers present in the Launch Control Centre, the Mission Control Centre and the VVIP gallery, I occupied my hot seat at the mission console next to Ramakrishnan.

Sitting there, I had a strange feeling. Memories of those two failures still haunted me. However, I told myself that I was fortunate to be given this opportunity to attempt this one more time. I recalled a couple of lines from the *Bhagavad Gita*. 'Handle the situation with equanimity, renounce all egocentric attachments, don't worry about the success or failure and act with equipoise in all situations.'

The countdown progressed smoothly without any hiccup or hold-up. About fifteen minutes before the scheduled lift-off, the automatic launch sequence was initiated by the mission director, Sivan, and the launcher was handed over to the mission computer for control of operations through the automatic launch sequence programme. Excitement and anxiety gripped everyone seated in the Mission Control Centre.

At the stroke of T_0, the stipulated time for lift-off, the four L40s came up blazing with all their might and GSLV-D5 lifted off from the second launch pad in full glory. The reverberation of GSLV drowned in the rapturous applause by our engineers at MCC and LCC and from those seated in the VVIP gallery. The excitement and joy was much greater than at the successful launch of PSLV-C25 that had carried the Mars Orbiter Mission two months earlier.

It was a bright afternoon and GSLV-D5 was visible all through its flight till the first stage. We had a great sense of satisfaction as the flight crossed the forty-eight-second mark, where the last mission (GSLV-F06) had failed. Soon the L40s and core solid stage S139 completed their burn and the liquid second stage took over easily. Within moments, the range operations director confirmed that the burn process in the second stage had gone normally. Soon the heat shield got separated and we knew the moment was just around the corner—the ignition of the cryogenic stage was going to take place.

'Cryogenic stage . . .' this time the voice of the range operations director got submerged under another round of thunderous applause. Yes! Our cryo stage had ignited and sustained the ignition. Some of our engineers could not suppress their joy and jumped off their seats, others threw their arms in the air in jubilation. The applause continued, covering all subsequent announcements confirming the success of the flight. This was in stark contrast to the normally reticent demeanour of our engineers during a launch.

My colleagues—Ramakrishnan, Prasad, Chandradathan, and Shivakumar—were visibly happy. I too felt goosebumps. This was our moment, this was the moment which we had strived for and the success was within sniffing distance. I did my best to hold back my emotions but it was difficult.

One thousand and twelve seconds after the lift-off, and after a textbook performance of the Indian cryogenic stage, GSLV-D5 pushed GSAT-14 into its intended orbit with an inertial velocity of close to 9.8 kilometres per second. We had won the battle.

I stood up and congratulated Ramakrishnan who had played a major role in this mission—first as the director of LPSC and then as the director of VSSC. We hugged each other with a deep sense of accomplishment. I congratulated all my colleagues around and of course the duo—Sivan and Vishnu Kartha. While many were visibly happy, many quietly shed tears. We had all tasted an extremely hard-earned success and there was no way we could hold back.

I made some initial comments on national television and then left the podium for my colleagues. It was the success of the team; and they had earned the right to address the nation through the media that stood firmly behind us.

I thanked all and truly felt we had repaid the faith that the country had bestowed on us.

26

MARTIAN ODYSSEY

The mystery of the universe, the galaxies, the solar system, with its planets and their moons, has mesmerized humankind since time immemorial. There are fundamental questions on the origin, form, content and future of these creations and the evolution of life on planet earth. Where do we come from? Where are we going to? Are we alone in the universe? The pursuit for answers continues. As we understand some of it, more questions surface and the saga of human inquisitiveness continues. The father of the Indian space programme, Dr Vikram Sarabhai's vision of establishing application of space technology as the mainstay of the Indian space programme was well implemented through the decades. Space application still remains the most important portfolio of ISRO. Still, there was scope for missions like the Space Capsule Recovery Experiment, Chandrayaan-1 and the proposed Human Spaceflight programme. These advanced missions called for the development of several new technologies which, in turn, could be brought back to the operational space programme to benefit the common man and society.

Reaching Out to Our Celestial Neighbour

The Chandrayaan-1 mission was successful in many ways in meeting these objectives. This was the first time we took a spacecraft out of the earth's orbit, and we made it traverse almost 4,00,000 kilometres to orbit another body in space. This was a long distance compared to our geostationary satellites which operate from an altitude of a mere 36,000 kilometres from the equator.

Of course, the Chandrayaan-1 mission enabled the detection of water molecules on the lunar surface and that brought a lot of joy and pride for the country, but this was not the sole achievement of the mission. We developed new technologies, developed a new version of PSLV, understood the mission strategies to undertake deep space missions and created the Indian Deep Space Network with a thirty-two-metre antenna at Byalalu near Bengaluru, and also established an Indian Space Science Data Centre at the same place. More importantly, Chandrayaan-1 created a belief in us that we could achieve things that NASA or ESA had. That was a very important feeling for the country in broader terms.

We could not afford to rest on our laurels after Chandrayaan-1. So we kept studying and reviewing our options. And then the dictum came, in 2007, from ISRO's Advisory Committee for Space Science (ADCOS—a group of senior space scientists of the country, led by Prof. U.R. Rao) that the next stop was Mars, the red planet.

There are a lot of interesting facts about our nearest planetary neighbour. Mars is just 55 million kilometres away from earth when it is closest, while at other times it could be up to 400 million kilometres away. It receives nearly 45 per cent of sunlight compared to the earth and has a gravity which is nearly 40 per cent of that of earth. Its diameter is half that of earth. Its atmosphere is rich in carbon dioxide. Fascinatingly, scientific evidence from ground-based observations and space probes suggests that Mars once had rivers, streams, lakes, and even an ocean; their discovery of the presence of methane on Mars suggests that life once existed or could have existed on this red planet. Mars, like earth, has an atmosphere, water, ice and geological features that all interact with each other to produce the dynamic Martian environment. It is abundant in iron oxide which prominently gives it the red colour. Our progeny might even get there to learn and live!

Interestingly, the ancient Indian astronomical text *Surya Siddhanta* provides an estimation of the diameter of Mars and the distance from earth. As a child, I was enamoured by a musical composition by Muthuswamy Deekshitar (1775–1835) where Mars has been described as *'Bhoomikumaram'* (son of earth) and *'Dharaneepradam'* (giver of land), etc. These are suggestive of the intense engagement of ancient India in astronomy, astrophysics and planetary studies. It must have been an astonishing intellectual exercise for our predecessors.

Like many of my fellow space professionals, I too was enamoured by the idea of an interplanetary probe to Mars while I was the director of VSSC. The possibility of an Indian Mars probe had been discussed in our weekend sessions with V. Adimurthy, the then associate director of VSSC, and his brilliant engineers led by R.V. Ramanan.

After taking over as the chief of ISRO in November 2009, I wished to pursue this with passion and professionalism. Adimurthy superannuated in May 2010, and I did not lose a minute in inviting him to the headquarters as a Satish Dhawan professor to contribute as our adviser on the interplanetary mission. Within three weeks of his taking over, we had a deliberation involving all centre directors of ISRO and the director of the Physical Research Laboratory at Antariksh Bhavan. We discussed the possibilities and constraints of a spacecraft mission to Mars that could just fly by the planet for a short while or be manoeuvred to orbit Mars several times (it was too premature then to even think of a lander mission to Mars). The consensus was that we attempt an orbiter probe at the earliest possible opportunity, based on the favourable earth-Mars-sun geometry.

We knew that the fundamental constraint was the capacity of the launchers that we possessed. Our versatile and reliable PSLV-XL could loft about 1400 kilograms into the heliocentric Mars transit orbit. In comparison, the Atlas V launcher of the

US could carry ten times heavier payloads to the same orbit. During the mission studies, we realized that with the current launch capability we were not in a position to deliver a satellite of meaningful mass straight to the Mars transfer trajectory exactly as NASA was planning with their MAVEN spacecraft on board an Atlas V launcher. We had to adopt a novel approach to reach that point with six orbit-raising manoeuvres from where PSLV-XL was to deliver our spacecraft. In this approach, we had to impart an additional velocity of about 1.5 kilometres per second by burning propellants packed in the spacecraft itself to reach a point from where it could get into the trajectory towards Mars. Earlier, we had manoeuvred Chandrayaan-1 to traverse 4,00,000 kilometres from earth and capture an orbit around the moon. We had to upscale the previous exercise for a voyage 1600 times longer. All of us were clear that our first mission to Mars should primarily aim to establish our technological capabilities to reach there.

By the first week of August 2010, I set up an interdisciplinary feasibility study team comprising seventeen scientists and engineers, and Adimurthy as the chairman. Nearly 125 experts from all centres and units of ISRO and PRL contributed to the exercise. Radhakrishnan Durairaj from the Launch Vehicle Programme Office of the headquarters ably coordinated the detailed study that spanned from August 2010 up to June 2011. In the meantime, the senior functionaries from ISRO centres and units were sensitized about this initiative in a meeting held in October 2010. All were excited to ponder over this new dimension and its technological challenges.

Finally, on the forenoon of 8 June 2011, Adimurthy and Radhakrishnan presented the feasibility report to me. In a nutshell, the study revealed feasibility to avail the next available opportunity in November 2013 for the mission. Also, it was clear that a spacecraft weighing 1350 kilograms (including 850 kilograms of propellant) would do the job and it could accommodate up to 25 kilograms of scientific payloads too. The study had delineated the complexities of navigating a spacecraft to Mars, and a strategy to use our versatile and reliable PSLV-XL launcher to place such a spacecraft into an orbit around earth. It had also listed satisfactory conditions for minimum energy transfer to a Martian orbit. With the fuel on board, we could place the spacecraft into an elliptical orbit around Mars—an orbit where it would be as close as about 500 kilometres from Mars at the periapsis of the ellipse and as far as about 80,000 kilometres at its apoapsis. Traditionally, Mars had been a notorious destination with a 60 per cent failure rate of the fifty-one Mars missions executed till then by the Soviet Union/Russia, the US, Europe and Japan. Finally, the study indicated the kind of scientific experiments or observations that we could attempt if the spacecraft reached the Martian orbit successfully.

It was indeed a commendable and exhaustive study as mandated. I was excited and elated. I could not suppress my excitement and said, 'You made ISRO's day.' Both of them were stunned as they did not expect such spontaneity.

Three of us did an in-depth analysis, asking some of the obvious questions to ourselves. Would it be better to wait for the subsequent opportunities of early 2016 or mid-2018 for the launch? Would it be advantageous to use a more powerful GSLV launcher instead of the PSLV-XL? If so, could we pack the spacecraft with more propellants to capture an orbit closer to Mars or carry better scientific payloads? The answers were readily available in the study.

And then came the logical explanations. First, the minimum energy transfer from the earth's orbit to the Martian orbit could take place only if the launcher injected the spacecraft into the specified elliptical orbit satisfying a stringent parameter called the 'argument of perigee' (the position of the perigee in the orbital plane). In this case the 'argument of perigee' was required to be around 280 degrees. This was considered even with the minor variations that occur because of the day and time of the launch.

On paper, the PSLV-XL launcher was capable of doing this with some tweaking in its flight sequence. We calculated that the first three rocket stages would provide the nominal velocity of 7.73 kilometres per second, sufficient to take the spacecraft to a safe orbit. And then we would have to wait for about twenty minutes before igniting the final rocket stage to impart an additional 2.2 kilometres per second (which was within the capability of the fourth stage). According to our calculations, we were confident that we could meet all specified injection conditions including the required 'argument of perigee'.

It is true that GSLV had the ability to carry a heavier spacecraft of say 2000 kilograms. However, we were yet to test our cryogenic stage at that time. But the launcher, with its three-stage configuration, had a severe constraint. The incremental velocity imparted by the two lower stages was limited to 5 kilometres per second. It would have been grossly inadequate to permit a delayed ignition of the last stage—the cryogenic upper stage. Also, the GSLV could ensure an 'argument of perigee' of 180 degrees, as opposed to the required 280 degrees. This meant the spacecraft would use up an extra amount of propellant even before reaching the Martian transfer trajectory—the starting point for the cruise towards Mars. Even at a later stage, if we wanted to use the GSLV or LVM3 for the Mars missions, we had to modify them with a 'restart capability' for the cryogenic stages or add a new propulsion module. Considering these points, there was no way, at that point of time, we could have used GSLV or LVM3 for a mission to Mars.

The study also highlighted the fact that the forthcoming three opportunities—the first in November 2013 and the third in May 2018—were only possible if we selected PSLV-XL for the mission. For the second opportunity (early 2016), PSLV-XL would not be sufficient. The choices were crystal clear: either launch a 1350-kilogram spacecraft in November 2013 or wait for the launch of a similar spacecraft in May 2018; both had to be done by the PSLV-XL only. The question of waiting for the GSLV did not arise because of its shortcoming to achieve the required injection parameters. We concluded that 'time was of the essence' for this mission—either launch it by November 2013 or wait for another five years.

I had no ambiguity in choosing November 2013. It was a daunting task to fulfil, but it was worth venturing into with all the might and determination of 'team ISRO' at hand. Besides the technological and scientific gains, for such a mission would provide great strategic advantage to India, it would generate immense national pride and excitement among the youth.

But I needed to muster support from within the organization and obtain approvals from the government to undertake such a high-profile mission on such short notice. I decided to tackle all of them head-on.

I approached Prof. U.R. Rao and, said 'Sir, I am coming to meet you along with Adimurthy. There is something exciting I need to discuss.' As anticipated, Prof. Rao was visibly excited with the idea and said, 'We should do it in November 2013 itself.' Nothing else was expected from that adventurous hero of Aryabhata. Prof. Rao added, 'Rad, let us meet tomorrow, get Goswami (Prof. J.N. Goswami, the then director of Physical Research Laboratory, Ahmadabad), Navalgund and Alex. Let's see what science we can attempt.' At that moment, Prof. Rao enrolled himself as the 'youngest scientist' of the first Indian Mars mission!

Next on my list was Dr Rangan, the then member (science) of the Planning Commission. Dr Rangan was thrilled by this logical next step after Chandrayaan-1, and he was quick to extend his support. 'Rad, see whether the Jet Propulsion Laboratory (of the US) can fly any of their scientific payloads in this mission,' he advised.

Confident with the responses of these two former chairmen and the most celebrated space scientists of the country, I called up national security adviser, Shivshankar Menon. I also sent a brief note on this sterling opportunity that India could avail to the PMO. Their approval came the same evening!

At the Drawing Board

We got on with the job to translate the ideas into action plans on fast track. The distance of nearly 55 million to 400 million kilometres from earth to Mars posed several challenges for the spacecraft design and deep space ground stations. A signal from the spacecraft could take twenty minutes to reach the ground stations as against 1.5 seconds from a spacecraft around the moon. The spacecraft had to be 'smart'; it had to be designed with sufficient autonomy for taking decisions, sensing its own health, without always waiting for commands from the ground controllers in case of emergencies. A clever three-tier antenna design and highly sensitive receivers in the spacecraft were essential. Besides, the Indian deep space station near Bengaluru had to be upgraded so that it could communicate with the spacecraft even if it was at its farthest location or to precisely track its position. The spacecraft had to be navigated with a full understanding of the pressure exerted by radiation and the gravitational force not only from the sun but also from other planets. After the 660-million-kilometre journey, we had to land the spacecraft within 500 kilometres of Mars; an error bar of 50 kilometres on either side was allowed. Dr Charles Elachi, the then director of the Jet Propulsion Laboratory, had famously said in a bilateral meeting that the navigational challenge in a Mars mission was similar to hitting a golf ball from their Pasadena (California) campus for it to land in the courtyard of Antariksh Bhavan in Bengaluru. I think the example summarized the essence of the challenge at hand. In the last phase of the mission, the spacecraft had to be braked to a certain extent, so as to be captured by the Martian gravity for it to start orbiting around it. To effect the brake, we had to wake up the satellite propulsion system after its hiatus of nearly ten months since the last operation. And most importantly we had to develop the spacecraft and the scientific payloads within just two years.

As I said earlier, nothing excited ISRO more than a stiff challenge; synergy, coherence and collective resolve were at work across the organization. At Thiruvananthapuram, Veeraraghavan and Ramakrishnan, the centre directors of VSSC and LPSC, explored all avenues for stretching the payload capacity of PSLV-XL, mission sequences, etc. At Sriharikota, Chandradathan, Prasad and Seshagiri Rao prescribed the steps to undertake a launch even during the cyclonic season of October–November. At the ISRO Satellite Centre, Dr Alex and Annadurai (the famous project director of Chandrayaan-1, who had been elevated as programme director for scientific and remote sensing satellites) worked out details of spacecraft configuration and mission operation. Shivakumar and B.S. Chandrasekhar worked out the essential upgrading at the deep space station, support from global deep space stations, and the need for ship-borne tracking terminals in the South Pacific Ocean. The target of November 2013 did not look insurmountable but we had to keep tabs on the readiness of the spacecraft.

We did a thorough costing exercise. The total cost of developing the spacecraft, realizing the PSLV-XL launcher, ground segment and operational expenses, boiled down to Rs 450 crore. That was probably a contribution of less than Rs 4 by each citizen of India. It was affordable by any means.

Armed with the basic facts and practical details of the mission by August 2011, we commenced wider consultations to elicit comments and endorsements from the scientific fraternity in the country to take the mission forward.

Prof. M.G.K. Menon, Prof. Yash Pal, Dr R. Chidambaram, Dr Kasturirangan, Prof. Goverdhan Mehta and Prof. Roddam Narasimha were briefed and they came on board. At Bengaluru, presentations were made before Prof. U.R. Rao and Mr Madhavan Nair as well as ISRO's veteran directors, who continued to be active members in the affairs of ISRO even after their superannuation. Chairman, ADCOS, Prof. U.R. Rao, was fervent with more than thirty ideas of scientific experiments that were elicited from the Indian scientific fraternity. Finally, he shortlisted nine promising experiments (proposed by ISRO and PRL) on scientific themes of interest. However, we decided to make the final choice by April 2012 based on the maturity of development trials of the instruments.

Excellent observations and suggestions came from former chairman Mr Madhavan Nair and Mr N. Vedachalam, a former director of LPSC. Mr Madhavan Nair's observations pertained to orbital accuracy, thermal aspects and sharpening of science objectives. He also suggested advance actions to demonstrate the restart capability of the liquid engine and development of scientific payloads in view of the tight schedule. That was the wisdom he gleaned from his experience with the Chandrayaan-1 mission. Mr Vedachalam advocated a thorough review of the flight dynamics and exhorted us to simultaneously start working on the second Mars mission due in 2018. It was an all-round support system, and we welcomed the counsel to delineate mission risks and success criteria.

I was assured that I could count on the enthusiasm and wisdom of all these veterans of ISRO for this formidable mission. They had gone through several such formidable phases in their career, and had developed satellites and launchers practically from scratch.

The first milestone of the government's approval process was achieved in December 2011 when the Space Commission deliberated on all aspects of the mission. It was also considered that the Indian voyage to the red planet deserved to be proclaimed on Independence Day by none other than the prime minister during his speech from the ramparts of the Red Fort.

Incidentally, a disquieting development that affected Chandrayaan-2 was at our doorstep by May 2012. Chandrayaan-2 had been initiated in 2008 as a joint mission of India and Russia for in situ studies of the lunar surface. India was responsible for orbiter spacecraft and launch (by GSLV), while Russia offered to provide the Lunar Lander and Rover. By 2010, Russia restricted its participation to the Lander module. ISRO focused its full efforts to develop the Lunar Rover and set a target to launch Chandrayaan-2 by 2013–14.

As a consequence of the failure of the Phobos-Grunt mission of Russia (that also carried a Chinese Mars orbiter, Yinghuo-1), a series of efforts to increase the reliability of Russian planetary missions were set in motion. The Russians desired to fly their redesigned and heavier Lander module aboard the Russian Soyuz flights in an experimental mission in 2015 and a full-fledged science mission in 2017. The Russians did offer to fly the Indian Rover in their experimental mission of 2015, subject to limitations of mass and higher risk presumably inherent with their experimental mission.

The fallout of these programmatic realignments by the Russians turned out to be a blessing in disguise for us. Chandrayaan-2 was not to take-off with the Russian Lander at least for the next five years and there was no guarantee as to when we would get it later. We decided to take up development of the Lander module ourselves which was beneficial from a long-term perspective. Soon, the scope of Chandrayaan-2 was revised with an Indian Lander and Rover modules.

Announcing the National Mission

The conch was blown, announcing the Indian Mars Orbiter Mission on the sixty-sixth Independence Day of India, 15 August 2012, by Prime Minister Dr Manmohan Singh. The government had approved the proposed project cost of Rs 450 crores in toto and we promised to stay within it. It was a national mandate to ISRO; we had just fifteen months to build the spacecraft and launch it. We could not afford to let it slip.

The project kicked off and a formal management structure was put in place. The centre directors had their slice of the tasks cut out. Among them, VSSC was the lead centre for PSLV-XL with Kunhikrishnan as the mission director till the completion of the launch phase. ISAC was the lead centre for the realization of the spacecraft and mission management. Subbaiah Arunan was chosen as the project director for the spacecraft and Kesava Raju as the mission director of the post-launch phase. B.S. Chandrasekhar, director of ISRO's Tracking Network in Bengaluru, doubled up as the project director for the Deep Space Network and ground stations. While the final choice of the scientific payloads and science data utilization plan rested with ADCOS, we created a Payload Steering Group, chaired by Kiran Kumar (who had taken over as the director of SAC by then), to guide their engineering aspects and interfaces. Scientific secretary Koteswara Rao was entrusted with international interfaces, especially support from NASA and the Jet Propulsion Laboratory for tracking and navigating, using their global Deep Space Network. This was essential as our Deep Space Network was not sufficient to track and communicate with the spacecraft throughout its voyage.

The trans-Mars injection operation was to be done on 1 December 2013. Everyone realized that twenty-four hours in a day were not sufficient! What we witnessed was an unleashing of human energy at superhuman levels in order to make an impossible target possible.

Preparatory development activities of the past two years had brought sufficient clarity on the cardinal issues. We could bring in a few subsystems and components that already had flight heritage from our previous satellite missions. Also, we could borrow a few subsystems from Chandrayaan-2 in view of its impending delay of at least four years. There were still many new developmental tasks for the spacecraft communication system, propulsion and control systems, on board autonomy and scientific payloads, besides extensive testing of the spacecraft in a simulated space environment. The logics for autonomy in the spacecraft were programmed on the on board computers and these had to go through rigorous simulations in the laboratory. We took a no-waiver, no-risk approach at every decision point. New lessons in project management were learnt, practised and further evolved.

A standing scientific and technical review committee was set up to oversee the implementation and to iron out all technical interface issues between the launcher, spacecraft, ground segment and mission management. Dr T.K. Alex, who had just passed on the mantle of ISAC to Shivakumar, chaired this overarching committee. Considering the paramount importance of timely decision-making, I chaired a high-level review committee comprising all ISRO council members. We met once a month to closely monitor and facilitate the project activities. I was deeply involved in the technical activities and met the teams at ISRO Satellite Centre's laboratories, clean rooms and test facilities once a fortnight. Arunan's early morning SMS messages gave excellent summaries of the progress and problems of the previous night.

ISRO's normal prelaunch review mechanisms were put in place. The Mission Readiness Review Committee, co-chaired by Dr B.N. Suresh and Mr K. Narayana, had a unique mix of ten veteran directors and senior domain experts from both the past and current generations. The Launch Authorization Board, chaired by M.Y.S. Prasad, was entrusted to oversee the mission till the launch, and the Standing Spacecraft Authorization Board, chaired by Kiran Kumar, was to oversee the mission from launch up to Mars orbit capture and payload operations.

We assembled a group of Indian space technologists and space scientists on a common platform at Antariksh Bhavan for two days to take stock of the readiness of each segment of the mission: concerns, interfaces, reconfirmations, contingency plans, etc. The state of preparedness was endorsed. By the last week of September 2013, after a thorough pre-shipment review (we do it for all satellites), the Mars Orbiter Spacecraft was ready for transportation from the ISRO Satellite Centre to Sriharikota for launch on board PSLV-C25. We could then target the launch for the last week of October 2013. India was on track.

The focus shifted to Sriharikota. But nature threw us a stiff challenge. A cyclonic system (Phailin) was approaching the coastlines of Sriharikota Island. We preserved and protected the completely integrated PSLV-C25 rocket and the Mars Orbiter Spacecraft sitting on top of it inside the Mobile Service Tower.

For the PSLV-C25 to meet the requirements of the 'argument of perigee' at the time of injection into the orbit, we tweaked the flight sequence by delaying the ignition of the fourth stage after the separation of the third stage. At this time, the rocket would be above the South Pacific Ocean (a range of nearly

17,000 kilometres from Sriharikota). This range was beyond the reach of our ISTRAC ground stations at Sriharikota, Port Blair as well as at Brunei and Indonesia. Hence we had tasked two ship-borne tracking terminals—SCI Yamuna (PSV) and SCI Nalanda (PSV)—to be positioned in the South Pacific Ocean at strategic locations to monitor ignition of the fourth stage and injection of the Mars Orbiter Spacecraft after the fourth stage of PSLV completed its task. ISRO, DRDO and the Shipping Corporation of India did their best for this first-of-a-kind expedition with sophisticated tracking systems with a stop at Fiji Islands. It was not really a smooth cruise. There were technical glitches in the tracking system in one of the ships. To top it all, there was a threat of a hurricane in the South Pacific which could put the ships in danger. We actually had to reschedule the launch, not because of a technical issue in the spacecraft or rocket, but to ensure that the ship-borne terminals were in place to track those two important events. I had many sittings with the Launch Authorization Board to synthesize relevant inputs from all teams and agencies concerned and work out contingency plans in case these ship-borne terminals did not reach within the specified timeline.

India Lives the Martian Dream

Finally, on the fifth anniversary of the launch of Chandrayaan-1, 22 October 2013, the declaration was made to launch PSLV-C25 with the Mars Orbiter Spacecraft on 5 November 2013. The lift-off time specified, based on the mission analyses for that day, was 1438 hours (IST). This was again based on the requirement with respect to the 'argument of perigee' at injection. We had a leeway of just five minutes for the launch; lest the steering programme loaded in the flight computer had to be redone or the launch had to be postponed by a day.

Finally, the launch day, 5 November 2013, arrived. Incidentally this was the first launch that we had scheduled in the month of November from Sriharikota, amidst the cyclonic season. Cyclone

'Phailin' had hit the coast a month ago and the next one, 'Helen', had not yet shown up. Our meteorology teams had reliable micro-level forecasts for winds, rain and thunderstorms that helped to plan the prelaunch operations better.

On the day of the launch, the VVIP gallery was full of dignitaries—minister of state in the PMO, principal secretary to the PM, US ambassador to India, Prof. U.R. Rao, Dr Rangan, Prof. Yash Pal, Mr R. Aravamudan and our veteran former directors were the notable ones.

A passionate Adimurthy was happy to get back to the commentary box one more time. He presented the proceedings like a live classroom lecture for the viewers in India and abroad. Mission director Kunhikrishnan authorized the automatic launch sequence computer to take over the control of the last ten minutes of operations and rigorous checking of the health of the PSLV-C25 and spacecraft. That was the twenty-fifth launch of the PSLV, and Kunhikrishnan was directing the mission for the ninth time. PSLV-C25 beautifully took off from the first launch pad, carrying the hopes and aspirations of 1.2 billion Indians to a destination where only three other nations had been before.

The first three stages performed perfectly, and then there was the expected communication blackout as the rocket went beyond the tracking range of our ground station. Everyone in the Mission Control Centre was getting restless for the confirmation from the ship-borne tracking terminals. The eagerly awaited signal came thirty-three minutes after lift-off and was welcomed with thunderous applause from the VVIP gallery. Though we never doubted our prediction of the flight path of the coasting phase of the PSLV, we still heaved a sigh of relief when the confirmation from the ship-borne terminals came in. The thin line between national fame and national shame was, as if, looming large. Ignition of the fourth stage took place just two minutes after the ship-borne terminals picked up the signal. Within moments we got the confirmation that the performance of the fourth stage was normal.

After the longest (till then) and probably the most important flight of PSLV, forty-three minutes after the lift-off, the Mars Orbiter

Spacecraft was precisely delivered into an elliptical orbit of nearly 250-kilometre perigee and 23,500-kilometre apogee around the earth, conforming to all the stringent specifications imposed on it. Thus the first of the three crucial milestones of the Mars Orbiter Mission was successfully achieved.

The fulcrum of action then moved from Sriharikota to the ISTRAC facility at Bengaluru, headed by B.S. Chandrasekhar, and mission director Kesava Raju took the cockpit. The Spacecraft Authorization Board, chaired by Kiran Kumar, kept a close watch along with his associates. I continued my deep involvement as a permanent invitee to their deliberations over the next year.

The team of 250 scientists and engineers, who swung into action at Bengaluru, was a wonderful mix of age and domain experience. Their enthusiasm was infectious as they had a new challenge at hand over the next eleven months. We saw many of the youngsters in their late twenties and early thirties coming up with novel ideas, truly exemplifying lateral thinking.

The spacecraft was in good health and ready for the next steps. The PSLV-C25 had injected the spacecraft with an imparted velocity of around 9800 metres per second. The objective of the subsequent operations was to raise the orbit of the spacecraft by imparting additional velocity of around 870 metres per second by firing the liquid engine in the spacecraft. The first orbit-raising manoeuvre of any spacecraft is a real test as its liquid engine (of 440 Newton thrust) gets initiated to impart the extra velocity while a set of eight tiny thrusters keep tabs on the correct attitudes of the spacecraft. The first orbit-raising manoeuvre for the Mars Orbiter Mission (MOM) was successfully executed on 7 November and the apogee of the spacecraft was raised to 28,825 kilometres; four more such orbit-raising operations were to be done. This was a golden opportunity for us to exercise and test out the built-in provisions for spacecraft autonomy.

The process went on smoothly for the second and third operations as well. The fourth operation on 11 November revealed one snag when the built-in redundancies for the solenoid coils of the fuel flow control valve were exercised. In one of the three planned contingency modes, i.e. when both primary and redundant coils were commanded to work in parallel, the fuel flow to the liquid engine stopped and the operation got aborted. We realized this and concluded that this parallel mode of operating the two coils was not possible for subsequent operations; they had to be operated independently in sequence henceforth. The very next day, we resumed orbit-raising operations as required.

The striking feature of these operations was the collective preparedness of the 250-member team for executing a complex sequence of operations and facing contingencies. The teams prepared a contingency plan for all possible eventualities during any of these operations. They thoroughly simulated those plans, tested and validated them, and conducted a briefing session to harmonize and familiarize all members of the team with their roles in such situations. Thus the teams got prepared to face any operational trouble with a few options of mitigation. In case of surprises from the heavens, we were mentally prepared and better equipped to handle it.

All orbit-raising operations were successful and we were ready for the second crucial milestone—the trans-Mars injection or 'TMI'. This was the crucial slingshot with which the spacecraft was given the boost velocity of nearly 650 metres per second to escape from earth's gravitational sphere of influence and set in course towards the red planet in the sun-centric trajectory. The task rested in the safe hands of mathematicians and specialists in flight dynamics. Over the past year, I had taken regular coaching from our flight dynamics expert, N. Gopinath, of the ISRO Satellite Centre on the intricacies of TMI to understand and contribute in the deliberations. It did help.

It is a well-known fact that Mars completes only 53 per cent of its circular orbit around the sun by the time earth completes a full revolution. For the spacecraft from the earth to reach Mars, it needed to travel in a direction towards the future position of Mars. In simple terms, we say that the spacecraft should leave the earth in a direction tangential to earth's orbit around the sun; travel roughly one half of an ellipse around the sun; then encounter Mars tangentially to its orbit.

On the decisive day of 1 December 2013, we gathered at the Mission Operations Complex (MOX) at ISTRAC. Commands were loaded in advance in the computers of the spacecraft to initiate firing of the liquid engine at 0049 hours (IST) as the spacecraft would then be at its perigee point of 270 kilometres, and terminate it after nearly twenty-two minutes. We had to hire a ground station located at Hartebeesthoek Radio Astronomy Observatory in South Africa to track the event as the spacecraft was expected to be out of our tracking range. Mission director Kesava Raju confirmed that the spacecraft's health was normal and all conditions were met for the TMI to start at 0049 hours as planned. However, the hours leading to the mission were tough to pass.

The 'T_0' appeared but there were absolutely no signs of the engine burn from the satellite! Everybody was getting restless and the whispers grew louder. Ramakrishnan and Chandradathan, the former and current directors of LPSC respectively, were visibly depressed presuming that their baby, the liquid engine of the spacecraft, had failed to act. Suddenly B.S. Chandrasekhar, the director of ISTRAC, broke the news that comforted everybody.

'There was a thunderstorm at the Hartebeesthoek station and both antenna terminals at the tracking station failed just a minute before the start of TMI. The engineers will restore them quickly.' Suddenly the excitement returned at MOX, pushing some out of their seats. Much to our relief, the data from the Hartebeesthoek tracking station resumed. It showed that the TMI had been going on smoothly as per the commands. Within an hour, the mission director confirmed that the spacecraft was on course. During the post-operation review, he gave a stunning account of MOM's journey for the next four days—at 1630 (IST) hours, the next day, MOM was expected to cross one lunar distance of 3,84,490 kilometres. By 0216 hours of the third day (4 December), it would exit earth's gravitational sphere of influence—a distance of almost a million kilometres. With the successful completion of TMI, we had completed two–thirds of the mission's objectives.

This operation was truly a crucial one, as the domain experts across the world started believing that India really had a great chance of reaching Mars in its very first attempt, an unparalleled feat for any country till then. The European Space Agency (ESA) had been successful in its first attempt of 2 June 2003 with 'Mars Express' mission too. But, there were not many who believed that ISRO could achieve the same. We got some breathing time and space at that juncture as MOM was set for a nine-month cruise with gravitational force on its sail. All we had to do was to keep a close eye on the spacecraft's health and, if required, execute a few small firings for midcourse adjustments. If everything went according to plan, we only had the major operation of the Mars Orbit Insertion Manoeuvre coming up in the second half of September 2014.

Since the announcement of the mission, several primordial questions had been posed to us: Why should India spend precious money for the Mars mission when the country faces several other pressing problems like poverty, hunger and lack of sanitation on earth itself? We, at ISRO, had been facing such questions since we started our activities in the 1970s. We kept repeating the benefits of space research till our critics were convinced. But, just for those who like the numbers—India always had the lowest space budget to GDP ratio among the major space-faring nations. So, it was not fair to accuse our space programme of usurping hefty resources.

Being the national space agency, we had to respond to technical and genuinely inquisitive questions. We did that with the utmost sense of accountability.

However, there were also questions posed to us regarding our motive to hurry the mission or push Chandrayaan-2 on to the back-burner or postpone the GSLV indefinitely. We were also questioned about why we did not wait for GSLV to get ready to attempt a more meaningful mission. Sadly, some of these questions were raised publicly by scientists of international repute, who had been part of the decision-making process.

It is conceivable that any game-changing mission like MOM would encounter various perceptions and criticisms. We decided to march on with our pre-project deliberations and convictions. But, we did focus on the constructive criticisms and took them in our stride. In a sense, we really admired the sceptics and critics who actually helped us work better as a team and kept us on our toes to avoid complacency. We had to respond to their concerns in the best possible way.

First, we had to keep reminding the country at several platforms and forums that the Indian space programme itself cost less than Rs 50 per head every year. But, the direct and indirect benefits reaped every year were tremendously important and of much higher economic value. And we had a golden opportunity to prove that. Barely three weeks before the launch of the Mars Orbiter Spacecraft, an extremely severe cyclonic storm 'Phailin' made its landfall (on 12 October 2013), close to Gopalpur in Odisha. It wreaked havoc in parts of Odisha, Andhra Pradesh and Bihar.

The government evacuated nearly 5,50,000 people from the coastline of Odisha and Andhra Pradesh based on the reliable early warnings of the India Meteorology Department (IMD). IMD was able to issue the warnings based on ISRO's satellite data. This natural calamity and ISRO's response to it somehow reduced the criticism of the 'extravagance in space'. We reiterated, at every possible forum, that the Indian Mars mission would cost only Rs 4 per head for any Indian citizen to whom ISRO was accountable. This accountability to the taxpayer necessitated us to communicate, convince and carry them through the complex navigation towards Mars, educating them of the real challenges that we encounter every now and then.

We encouraged and facilitated renowned science communicators to write books on Mars exploration that would have educative value. Srinivas Laxman, Ajey Lele and the Pallava Bagla–Subhadra Menon duo did an excellent job. During August–October 2013, I engaged in one-to-one sessions with nearly thirty senior journalists at Antariksh Bhavan. Interviews on TV channels followed. We invited media persons from all over the country to a conducted tour and media briefing at the ISRO Satellite Centre while the spacecraft was in its final testing phase in the clean room. We repeated the exercise at Sriharikota where the PSLV-C25 was in its final phase of checking at the launch pad.

We sensed that we were not getting across to those who wanted to hear us—the younger generation of India. We had to establish direct communication with the younger generation with a distinct demographic profile—students, teenagers and young professionals, which meant people who had a stake in the development of science and technology in the country. We felt social media would be the best way to reach them.

In 2012, social media for a government organization was not popular yet. Concerned authorities had issued a fifteen-page directive which in some way or the other was difficult to implement. This was because of an older directive that prohibited government employees from having social media profiles or allowing social media tools to be used in government premises. We took diligent steps, followed the right mechanisms, and in less than four days between decision and implementation began to use social media through Facebook and Twitter. We created an in-house team of just three engineers to create and post content on behalf of the organization. ISRO was probably one of the first government organizations in India to start interacting with netizens proactively. The response was overwhelming, even by conservative estimates. More importantly, our social media tools became a platform not only for providing quick mission updates but for serious discussions on aerospace, astronomy and the Indian space programme as a whole.

Transparency, accountability and communication galvanized the country to journey with us. ISRO's Mars Mission did become every Indian's Mars Mission. I saw the pendulum slowly shifting from criticism to optimism. Nothing could have been more satisfying.

Meanwhile we had also completed the GSLV-D5/GSAT-14 mission with the Indian cryogenic stage which was a huge success after those two agonizing failures of 2010. ISRO was back in its pristine glory. Against the backdrop of this euphoria, we inched towards the next big moment—the Mars Orbit Insertion (MOI) of MOM. D-Day was calculated to be 24 September 2014. The Standing Spacecraft Authorization Board, under its chairman Kiran Kumar, and the entire mission team of 250 engineers and scientists had six sessions at different intervals for a full day each. I attended all the reviews with the rapt attention of a serious student.

ISRO Scripts History

We had clarity on a few basics. Our prime goal was to orbit the spacecraft safely around Mars. The liquid engine that had been sleeping since 1 December 2013 was the single crucial factor to reduce the velocity enough for the spacecraft to be captured by the Martian gravity. It was imperative that we confirmed the condition of the engine well in advance.

The contingency plan was to collectively fire the eight tiny thrusters meant for attitude correction to brake the spacecraft. It was also imperative to judiciously utilize every gram of the fuel and oxidizer to maximize our probability of success in Mars capture. We faced the constraint of drawing the fuel and oxidizer in 'blow down mode' as the pressurization system had been plugged in after TMI for safety reasons.

All operations were simulated on a spacecraft model and propulsion module mimicking their state during MOI. There was another issue with the communication delay of as much as 12.5 minutes one-way because of the galactic distance involved. The sequence of operations had to reserve adequate provisions for that. An intermittent communication blackout was also expected due to reorientation of the spacecraft during MOI operations. The spacecraft would also have to go through power starvation as it would go behind Mars where it would not receive solar power. We had to upload all the commands required for MOI into the spacecraft ten days in advance to guard against any eventualities that may occur at the ground tracking stations. The silver lining for us in this process was that we were dealing with a spacecraft in perfect health.

On 12 September, the Spacecraft Authorization Board formally cleared the action plans and contingency plans for the next twelve days. They authorized mission director Kesava Raju to go ahead with the necessary rehearsals and mission operations. The same afternoon, I addressed the entire ISRO team from the auditorium of the ISRO Satellite Centre. I congratulated everyone for contributing to this mission and explained the course of actions, the precautions and contingency plans.

On 22 September, the spacecraft entered Mars's sphere of influence. We decided to activate the liquid engine for about three seconds for the final correction of trajectory. We decided to use the main thrusters instead of using the tiny thrusters tasked for that role. If the liquid engine restarted on that day, we knew our intended mission would go as planned. Otherwise, we would have to resort to plan B of roping in the eight tiny thrusters which could at best burn for about ninety minutes and end up probably orbiting around Mars once, that too a few hundred thousand kilometres away from it.

Hence, 22 September became the de facto D-Day for the mission executives and ISRO teams. At the Mission Operation Complex, we were joined by Prof. U.R. Rao and Dr Rangan who could not keep away from this historic moment. They sat alongside me for the operation. The stone-faced Kesava Raju, the mission director, was sitting still in his chair, wearing a look of confidence and conviction. All of us in MOX eagerly awaited confirmation of the three seconds of action in the spacecraft that would reach the earth after a delay of ten minutes.

This time there was no interruption, no natural calamity and of course no malfunctioning of the ground system. We got the confirmation that the guns were indeed blazing and the hiatus of nine months of non-operation could not dampen it. At that moment, we could say with 90 per cent confidence that ISRO was going to create history two days later. But, we decided to keep it to ourselves and kept our fingers crossed as our actions of the next two days had to go well.

I had been keeping the prime minister periodically apprised of the preparations for the historic event and had also proposed to set up necessary technical facilities at '7RCR' for the PM to witness the event live and address the nation from there. Soon after my update on the possibilities of a plan B, it was hinted that the PM intended to be with us at Bengaluru on that morning. That remarkable gesture of leadership from the PM electrified the entire team and it gave me some comfort for the encounter with Mars. The PM and his entourage arrived at Bengaluru on the previous day to join us in the early morning at the Mission Operation Complex.

Rashtrapati Bhavan was also briefed on the sequence of events and we were assured that the President of India would certainly witness the event.

I made it a point to make a courtesy call to former PM Dr Manmohan Singh and apprise him of the impending finish of the historical journey that he had approved two years ago. He wished me and the team the very best and expressed confidence that we would bring glory to the nation.

On 23 September, a day before the D-day, we had a very pleasant surprise—Kalam Sir decided to take a detour from his Chennai–Delhi trip and join us at Bengaluru. He spent a couple of hours at ISTRAC, greeted everyone present there and listened to a briefing by mission director Kesava Raju at MOX. Kalam Sir, our first mission director of SLV-3 in 1979–80, appeared satisfied with our preparations. I could guess that he was in two minds, whether to stay back with us for the next day or honour his commitment for a convocation address at a university in North India. With childlike reluctance, he left for the airport to catch the last flight to Delhi but reminded me to keep him posted on the progress because he wanted to mention it during the convocation address the next afternoon.

The next morning started quite early for us. For an operation that was to start at 6.55 a.m., we reached ISTRAC by 2 a.m. Some wished to stay back on the campus. Excitement, hope, anxiety—all emotions were at play on the campus. The presence of the PM and his security detail added more thrill to it. The seniors arrived and took their respective seats. The upper gallery of MOX was reserved for the PM's entourage. Upon their arrival, I received the PM along with the director of ISTRAC. After the customary exchange of pleasantries, I requested the PM's permission to sit with my team and proposed that Dr Alex brief him during the proceedings. The PM happily agreed and wished me luck.

I returned to my seat. Everybody took their seats except for Arunan (the spacecraft project director), who would never sit at his designated place inside MOX but preferred to walk outside MOX every time there was an operation. Soon we were given the confirmation that the liquid engine had started firing albeit after a communication delay of almost twelve minutes. Within minutes, the spacecraft was in the communication blackout zone as the red planet came in between MOM and the earth. At 7.41 a.m., the burn was expected to stop after effecting the requisite braking. The huge display board did not show what was happening. Although the blackout was over in the next seven minutes, we had to wait longer due to the communication delay. After an unbearable nineteen minutes, we got the confirmation that the braking process had been successful and based on the calculation, we knew we were orbiting Mars. We had created history. More importantly, ISRO had retained its legacy. The entire MOX erupted with joy and cheers. I am sure so did all our fellow Indians who were glued to their TV sets since early morning.

A formal confirmation call came from the mission director, who had been directed not to announce the result. We wanted the PM to announce this to the nation. I slowly walked up to him and he, having already guessed the result from the spontaneous reaction of the engineers at MOX, gave me the warmest hug I had ever received. There were both cheers and tears as our engineers went around congratulating each other.

For all our fellow Indians, 24 September began on an altogether different note as the PM started his address saying, 'Aaj Mangal ko MOM mil gayee . . . Ma kabhi niraash nahi karti [Today Mars got its MOM . . . A mother never disappoints].' The spirited address by the PM was heard with rapt attention by the global community. In his hour-long speech, the PM passionately praised ISRO's culture and scientific temper. We felt that this was bigger than any certificate from anybody for the effort that we took. Those few sentences erased the tiredness of the numerous sleepless nights and the exhausting hours in the laboratories.

Indians around the world watched with exhilaration those spectacular moments of Indian history enacted in the Mission Operation Complex on the morning of 24 September 2014. The world looked on in awe, the country rejoiced, and it was an inspiring moment for the children assembled in their schools to follow the Mars mission.

After the speech, the PM went around MOX and interacted with the more than 300 scientists and engineers present. He gave several autographs to our young colleagues and shared a word or two with everybody. The PM then joined us for a cup of tea in the adjoining room along with several of his cabinet colleagues as well as the governor and CM of Karnataka. During tea, we heard the announcement from the console of the mission director, 'The imaging camera aboard the spacecraft has taken the first photographs of Mars.' The day had started for MOM and she indeed got busy with her work. Just before the PM boarded his aircraft for Delhi, I was able to inform him that the first photographs of Mars had been processed and they were stunning.

The PM wished to see these pictures and asked me to send them to him the next morning at '7RCR'.

The President of India, Pranab Mukherjee, called me up personally to convey his hearty congratulations.

Congratulatory messages poured in and we were touched by the heap of postcards that we received from schoolchildren from different parts of the country.

The Indian Mars Mission would probably go down as one of the most memorable feats in the history of science and technology in India. Not only had we succeeded in reaching the red planet on the first attempt; in the process we had developed a number of technologies that could help us better our operational space systems in the coming years. More importantly, we had instilled a belief in the country's youth that space technology was an area where we were on a par with the world's best. We also proved nothing was impossible given the ability and resolve to do it.

The rest, they say, is pleasant history. ISRO did a mission to Mars in just four years between the feasibility study and fruition. On 19 October 2014, Comet Siding Spring passed by Mars and we tweaked the spacecraft's orbit to hide behind Mars to avert any harmful impact from the comet's tail dust. Thanks to the autonomy built into the spacecraft, it successfully survived a solar conjunction during 27 May to 1 July 2015, when it had no contact with any stations on earth.

As I write my memoir, MOM continues to run, beyond her projected operational life of six months and is multitasking, with her instruments—Thermal Infrared Imaging Spectrometer, Methane Sensor for Mars, Lyman Alpha Photometer and Mars Exospheric Neutral Composition Analyser, besides beaming the stunning photographs shot by her Mars Colour Camera.

The PM had rightly said, and we do believe that *Mom never disappoints*.

27

THE NEXT FRONTIER: LVM3
AND HUMAN SPACEFLIGHT

'The journey of a thousand miles starts with a small step . . .'

Since the early 1990s, India had occupied a pre-eminent position in the world by virtue of its capability of building its own spacecraft, launching them from its own spaceport on board indigenous launch vehicles and operating them for the welfare of the nation. This combination of capabilities had placed India in a select group of six nations—the space-faring elite of the world. Still, there remained two more summits to scale.

We are still dependent on foreign launch vehicles to orbit our heavier communication satellites (more than 2.5 tonnes of launch mass), and, of course, we are yet to establish our presence in the realm of manned spaceflights. Two of my illustrious predecessors had initiated programmes in those advanced areas. When I took over at the Vikram Sarabhai Space Centre in 2007, these automatically became my priorities, continuing later as I became the chief of the Indian space programme.

The first of the initiatives was the development of India's advanced launcher LVM3 with a twofold capacity of GSLV, which Dr Rangan had taken up in 2002. The second one was Mr Madhavan Nair's Human Spaceflight initiative of 2004. Both the initiatives called for a plethora of developments in the coming years.

LVM3 had been configured with less propulsive stages compared to PSLV and GSLV as well as their legacy and learning. Of course, LVM3 was a giant, almost twice as heavy as PSLV but it could deliver 3.5 times heavier satellites into similar orbits. While LVM3 is nearly 1.5 times heavier than GSLV, it has the capability to deliver two times heavier satellites into similar orbits. Thus, LVM3 is bound to be a more reliable and cost-effective launcher in the future. Anatomically, LVM3 had two solid booster motors (S200) as the strap-on stages to the core liquid motor (L110). The S200 motors, having more than 200 tonnes of solid propellant each, were the third-largest solid rockets after the ones used in Space Shuttle (of the US) and Ariane-5 (of Europe). The L110 stage was derived after upgrading and clustering two Vikas engines that serve in the second stages of PSLV and GSLV. On top of these two stages, LVM3 had a high-thrust cryogenic stage (C25) that was meant to deliver almost 50 per cent of the incremental velocity to the payload.

In the development of LVM3, we had crossed two significant milestones. We had qualified the core L110 stage through 240 seconds of test-firing in one go at Mahendragiri during March–June 2010. Secondly, we completed the static test of the S200 solid strap-on motors in January 2010 and required one more similar test to qualify S200 for flying on LVM3. However, there were issues to be resolved in the development cycle of the cryogenic engine (CE-20) meant for the C25 stage. Unfortunately for us, the consequences of the prolonged recovery process of the cryogenic upper stage of GSLV caused the slowdown in the development of CE-20 as the resources were being shared between the two programmes.

Meanwhile, a closer look at the configuration of LVM3 revealed that its ascent through the crucial atmospheric phase was trickier to tackle, primarily due to the inherent higher aerodynamic instability. An aerodynamic coefficient μ_α about the pitch axis of LVM3 would be around 3.5 times higher than that of our PSLV-XL. That implied that the autopilot of the launcher had to be extremely agile to compensate the build-up of errors in attitude. We had done a lot of wind tunnel tests and mathematical modelling to understand the boundary conditions. But, these empirical analyses had limitations. The efficacy of the control systems and the practical ramifications of loads on the rocket at critical events are best confirmed through a flight test. Further, we required a clear picture of the complex flow fields between the fairly large heat shield and the lower stages.

I had been toying with the idea that, in the given settings, the best strategy for ISRO was to take up an experimental flight of LVM3, so that we could check all such issues and correct them if necessary. But I kept my fingers crossed for the development of the C25 cryogenic stage built around the CE-20 cryogenic engine.

I broached this idea with the project director of LVM3, Somanath. He had taken over the baton in June 2010 from N. Narayana Moorthy and had earned a place of pride on his own merit. There were perceptions that trusting a forty-seven-year-old engineer to steer such a high-profile project was an indication of my romance with the idea of empowering the youth; but I was convinced about his abilities. He had been with LVM3 for the past eight years. As the director of VSSC, I had been a beneficiary of his immaculate technical judgements. In the case of this idea of an experimental flight of LVM3, his nod was important for me.

Somanath grabbed the idea with both hands, and said that he had a list of a dozen more issues to be confirmed if such an experimental flight was given the go-ahead. His concerns pertained to dynamics at lift-off; new flight software for guidance and control; dynamics of separation of the heat shield and the overall mission sequence; and so on. Hence, we could not delay the experimental mission further. However the development of the cryogenic engine (CE-20) and stage (C25) remained a question mark. If it was ready before the experimental mission, CE-20 would also be a part of it. Otherwise, after the ground tests it would be a straight fit into the LVM3, which would have qualified all lower stages through the experimental mission. We were also conscious of the fact that truncated test flights were in vogue elsewhere while developing heavy launchers. But, at ISRO, we were yet to try it.

I discussed the idea with Ramakrishnan, the first project director of LVM3 and the then director of LPSC. He smiled in endorsement and said, 'Nothing like it, if we could do so.' Veeraraghavan, director of VSSC, the lead centre for LVM3, endorsed the idea, as did Dr Suresh. Somanath was given a green signal to work out the details and present the idea at the ISRO council meet in October 2010, to all the senior functionaries at Antariksh Bhavan. The response was encouraging, to say the least.

By June 2011, the LVM3 team worked out definitive plans for the experimental mission and named it the 'X Mission' with four possible options for the passive cryogenic stage. The director of VSSC, Veeraraghavan, tasked an Integrated Technical Review of LVM3 under the chairmanship of Dr Suresh, with twenty-five aerospace experts drawn from academic institutions, national aerospace agencies and another twenty-five veterans of ISRO.

On 15 June 2011, during the first meeting of this Integrated Technical Review of LVM3, Somanath made a scholarly presentation and earned the respect and admiration of everyone present in the 300-seater auditorium at VSSC. There were differing viewpoints expressed on the scope of the LVM3-X mission. The review process continued and, by December 2011, the proposal for the LVM3-X mission was accepted, and details of the flight configuration, especially the architecture of the passive cryogenic stage were chalked out; it was a diligent synthesis and all were on board with the idea.

The only task left was to look for a payload of nearly 3500 kilograms that could be test-flown aboard this experimental flight, the scope of which was limited to an altitude of 125 kilometres. The lower stages of LVM3 together were to impart a velocity of 5.4 kilometres per second to the payload. Naturally, we could not launch a satellite because of the limitations in the achievable altitude and incremental velocity. We were in a state of confusion as to what best could be flown aboard the flight considering the given boundary conditions.

ISRO had a long-standing aspiration to fly a two-member crew to an orbit around the earth and bring them back safely after a stay of a week in space. These were the broad mission objectives that we had set for ourselves under the Human Spaceflight Project—a dream that we had been chasing since 2004. GSLV had the capacity to orbit a crew module with two crew members aboard but we had to ensure a demanding reliability goal of 0.98 (i.e 2 failures in 100 flights). The mission would cost Rs 12,400 crore, so it was yet to get a green signal from Delhi. In early 2010, a strategy was finalized to attempt the essential tasks in a phased manner.

The first phase of two years prescribed the flight testing of the unmanned crew module system and its service module in an exclusive PSLV flight and we sought approval from the government for this section alone. However, the recurrent failures of GSLV poured cold water on this approach.

Meanwhile, the design for the crew module was nearing finalization. We grabbed the thin opportunity of flying it aboard the LVM3 experimental mission. In fact, this was a golden opportunity for testing the crew module developed as one of the critical technologies. The scope permitted the crew module to be exposed to nearly 70 per cent of its nominal orbital re-entry velocity. Also, we had the option to test its deceleration system and recovery protocol.

The LVM3 experimental mission came as a breath of fresh air to Unnikrishnan Nair, the project director of the Human Spaceflight programme, and his team who had been craving for the opportunity to take forward the development of critical technologies beyond what had been authorized during 2007–2010. But, it was such a formidable challenge to deliver a flyable crew module in a short time frame. Also the module had to carry an attitude control system to guard against the nominal disturbances at injection by LVM3 at an altitude of 125 kilometres. But the team behind Unnikrishnan Nair took the challenge and the module was christened the Crew Module Atmospheric Re-entry Experiment (CARE).

We obtained the necessary endorsement of the government for this detour. I vividly remember that meeting in Delhi in April 2012, when Ramakrishnan and his disciple, Somanath, mesmerized all the members of the Space Commission with their lifetime's insights on launcher technology and the potential impact of this experiment.

What followed then was a race against time by both the LVM3 and CARE teams. While the majority of the launch vehicle experts at ISRO were immersed in taming GSLV, the persuasive Somanath and Unnikrishnan ran their competitive race. Somanath's second hat as deputy director of VSSC of the Structures Entity (2011–13) helped them to mop up all resources to complete the voluminous structural testing for all articles.

Once GSLV left the scene with its astounding success in the first week of January 2014, LVM3-X mission and CARE came to the centre stage. The Mission Readiness Review Committee and Launch Authorization Board were put in place. A high-level review committee chaired by me took charge of pushing both teams to their limits. While realizing the passive cryogenic stage, Karthikesan and V. Narayanan's team made sure to conduct a few successful tests of short duration on the CE-20 cryogenic engine, promising its logical direction to a full-fledged flight of LVM3 within the next couple of years.

During a light-hearted conversation in 2014, I asked Somanath if there were issues in the experimental mission on which he was losing his sleep. He smiled and accepted that there were a few. He

pointed towards aerodynamic characterization and flight software
as the potential problems. Two special teams were constituted
immediately—led by K. Sivan and Kunhikrishnan by virtue of
their domain knowledge (Sivan was the deputy director of the
aeronautics entity in VSSC before becoming the project director
of GSLV; Kunhikrishnan was in Quality Assurance before getting
into the PSLV project) to take a critical look.

There were a few more sessions of the Comprehensive
Technical Review by national experts. We organized a combined
review for the LVM3-X mission and CARE in August 2014 at
Antariksh Bhavan. The launch mass of the CARE module with
the attitude control system was finalized at 3775 kilograms and the
necessary refinements in the LVM3-X mission were incorporated.

We started the launch campaign with great levels of
enthusiasm. Unnikrishnan's team was still racing against time to
conduct sea trials jointly with the Indian Coast Guard, rehearsing
the recovery of the CARE module that was expected to splash
down about 1600 kilometres from Sriharikota and 600 kilometres
from Port Blair.

Amidst these high-octane activities, with my last month at
the helm approaching, I had a different thought process. Though
initiated way back in 2011, executing such a high-profile space
mission during the last month of one's tenure is a tricky affair and
one has to navigate diligently and dispassionately. There was no
place for undue haste or over-cautiousness. The decision-making
had to factor in all concerns, particularly the perceptions of the
prospective successors. At the same time, one could neither allow
nor afford matters to drift away. I was convinced that the mission
director Somanath and the mission teams had done whatever
was humanly possible for a successful flight test of the LVM3-X
mission and CARE. I gave them my green signal.

Finally, the Mission Readiness Review Committee and
Launch Authorization Board cleared the launch to take place
on 18 December 2014, and lift-off was scheduled at 0930 hours
with a rather long launch window of two hours. The essential

consideration was that we should be able to recover the CARE module before nightfall.

This was probably one mission where everything was perfect. LVM3 took off majestically with blazing flames from both S200 solid rockets; two minutes later the L110 stage joined the action with its twin engines. Two solid rockets and two liquid engines worked in tandem for a few seconds and the S200 withdrew from the scene after its task, as planned. The L110 stage completed its job in three and a half minutes, and CARE was delivered neatly at 126 kilometres altitude. Within another fifteen minutes, we got a confirmation that the CARE module had splashed down safely. It was just within a ten-kilometre radius of the predicted splash point. That was the remarkable accuracy demonstrated by this mission—something creditable for both LVM3 and CARE teams.

The flight data was invaluable to make minor modifications later to the heat shield and nose-cone of LVM3 to enhance robustness of India's workhorse launcher in the immediate future. The baby step towards India's Human Spaceflight was promising as the next logical step in India's space exploration.

On 18 December 2014, it was indeed a proud moment for me as I signed off from the Mission Control Centre in letter and spirit. I passed on my launch-apron to my beloved colleague and friend, Somanath, who immediately wore it then and there. He demanded my autograph on the coat and I happily scrawled my signature across the collar!

Somanath had accompanied me from Thiruvanathapuram, on my request, as I was ascending to the hot seat at Antariksh Bhavan. It was a moment of pride for me when he addressed the nation on national television after the launch. He was followed by Unnikrishnan, with whom I have always shared a special, personal bond. Unni later emailed me a photograph of our colleagues capturing the crew module at sea. I sent an SMS in return to confirm that I would be at Ennore Port near Chennai to receive them and bring back the CARE module from the Bay of Bengal.

I thought this was probably a fitting finale for a career which saw me traversing through the tsunami of the deep sea to the red planet on the wings of our Mars Orbiter Mission. I was fortunate and equally proud to have seen it all.

28

WORDS OF WISDOM

NILANJAN ROUTH

'Are you sleeping or what?' In fact I was, and that very familiar voice woke me up. It was Dr Radhakrishnan, standing next to the aisle seat that I was occupying on our return flight from Delhi to Bengaluru.

It was a freezing winter morning in New Delhi, on 30 December 2014. We, the passengers, were stranded inside the parked aircraft on account of the dense fog engulfing the tarmac. The flight crew said it would take 'some time before taking off' and for obvious reasons they could not allow us to disembark.

We were returning after a day-long meeting on the Make in India programme that was held at Vigyan Bhavan, New Delhi, the day before. The meeting had gone smoothly. The efforts put in by ISRO for close to five decades towards indigenization of materials, production and processes were well recognized by the industry associations present at the sector-specific meeting. Further, our proposal of assigning the operational component of the space programme (operational launch vehicles and standard satellite systems and services) had been received well by the industry and Delhi.

I had always been amazed by the extraordinary energy levels of this sexagenarian, who never seemed to get tired. The Make in India workshop had started at 8 a.m. the day before and had gone on for nearly ten hours. This was followed by a gruelling drive of around ninety minutes back to our guest house at Dwarka. Yet, he was up and ready for a customary post-dinner chat with us—the colleagues accompanying him to this meet.

At times, these post-dinner chats would continue till late into the night irrespective of the return flight's timing the next day. It was great fun as the topics under discussion were diverse—ranging from our own programmatic priorities to the global scenarios in the domain of space technology and so on. Sometimes, we would share our opinions and experiences, even from our personal lives, and the discussions would turn spicy, rich and energizing. Most often, being the youngest member in the team, I had the luxury of quietly absorbing and enlightening myself. In some part of my mind, I had a queer feeling that this luxury would be over soon and this was probably the last session, which we had thoroughly enjoyed.

'Yes sir, I kind of dozed off,' I replied sheepishly.

'Do you mind?' I was a bit shocked as he indicated that I should move over to the next seat. I did so because the flight was quite empty and I had all three seats in that row to myself. He gladly settled into the constricted economy class seat beside me.

'Tell me, what have you learnt from yesterday's exercise?'

I was pretty used to that question. Since the time I was moved from Antrix to the science secretary's office, every meeting, every exercise I had been a part of was followed with that question. It was a test of reconfirmation for him in order to check if I had grabbed the point or not. I was given full liberty to express my views even when our ideas and opinions were starkly contrasting. I knew I had an ear listening to me even though my views were not always put into action, I felt that it was a fair process.

This was an easier question, I thought. I slowly started, 'Well to the best of my understanding, sir, Make in India is a very important initiative to promote and push forward manufacturing in twenty-five key industries, which includes Aerospace. So, we at ISRO should look forward to it.'

He smiled and continued, 'What is the big picture; what exactly can we expect in the long run?'

'Isn't it too early to build a perspective?' I asked apprehensively.

'No it is not! In fact, this is the right time. As a growing and evolving national space agency, ISRO has the mandate to manage the operational space system as well as to develop technology and applications for future programmes. Our Twelfth Five-Year Plan proposals included twenty-five launch vehicle missions, thirty-three satellite missions, and a number of challenging technology developments as well as facility establishments. The number of missions and complexities of future missions is only expected to rise further.

'This is just the right moment to offload the routine activities completely to the industry. We had been deliberating the idea of augmenting the industry participation in the operational components of the Indian space programme like PSLV, the operational launch vehicles, geostationary satellites (GEOSATs) and navigational satellites or maybe satellite services.

'And let me tell you, this is not difficult. We have to start thinking in that direction somehow. The commercial potential of the operational component of the space programme, as we all know, is huge. Maybe it's time that we conceptualize an Indian equivalent of Lockheed Martin or Arianespace with the kind of support these organizations enjoy from their countries of origin. But the definition of the new organization and its operational principles need to be worked out in detail.'

I had a genuine doubt, 'How do you see ISRO's activities evolving over the period of time while the industry takes over the downstream activities?'

'There are lots of activities. Do you remember the prime minister's speech at SHAR after the launch of PSLV-C23 in June this year? I think we need to regroup our entire basket of Space

Application programmes to suit the vision and that target of the PM. As the chief minister of Gujarat, he had depended heavily on space-based inputs. It is natural to expect that the PM will have a definite role for ISRO in his scheme of things. In fact, if you remember, we were planning to conduct a two-day workshop on Space Applications involving all ministries of the Central government. ISRO is again at the crossroads of making itself highly relevant to national development.

'Having said that, ISRO must not look away from its own technology objectives. I am assuming the industry would be taking over the lion's share of ISRO's operational burden in three–four years' time frame. ISRO should chalk out its action plan for that improved national scenario. It's probably time to reconsider the 'Indian Human Spaceflight' initiative. And I am not saying this because I had put a lot of effort into it; I am saying this because this is the next logical imperative of our space programme. I am not saying India should have an ambitious Human Spaceflight programme, but we must evolve a mission strategy with the right mix of space robotics and human cognitive abilities for space exploration. We should be open to international collaboration for areas like astronaut training, space biomechanics and so on. We must attempt it in earnest.'

'It's indeed an exciting future you are talking about,' I said in appreciation.

'Yes and you are going to witness the excitement unfurl before you. But always remember that as a professional you should always be optimistic and never discard an opinion at face value. Remember the possibilities are endless and so are the abilities of our colleagues,' replied the veteran of ISRO.

I sensed that it was an opening and the right time to extract a few more words of wisdom. I inquired, 'Sir, how do people like us in the organization prepare for the big leap coming up and what do we do to ensure success?'

The flow of wisdom continued, 'There are two different aspects to your question. One is keeping pace with the evolution of the organization and then the most critical issue of success.

As I see it, we are fortunate to be a part of an organization that has been continuously evolving. A learning organization like ours gives ample opportunity for professionals to learn, implement and grow. That's a wonderful journey. In my case, I started as an avionics engineer, went on to pursue a degree in management, pursued doctoral studies and the organization supported me.

'I did my part and came back to the organization and had this wonderful opportunity to rediscover myself. Soon, I was called to the headquarters to handle Budget and Economic Analyses and then was handed another alien portfolio which I had to learn on the job. I don't know of many organizations which give so much opportunity to learn and deliver. And I don't think I did badly. I had the opportunity of working in so many diverse areas. Yes, I was lucky to be at the right place of work at the right time, but again, luck brings you the opportunity, you need to capitalize on that to take it further where it really matters.'

He asked for a bottle of water from a passing steward, paused to take a sip and then continued, 'Success is a tricky thing. It may or may not reflect the effort that you had put into it. The best thing is to concentrate on the tasks at hand. The results are not under your control so why lose sleep over it! But you need to prepare and prepare harder. I read somewhere that if you had nine hours to chop down a tree; you must take eight hours to sharpen your tool. Let me tell you, this had worked for me for years, and I am sure this would hold good for a professional in any domain.

'You know Nilanjan, there are things more important than talent, and hard work and luck—these are character and conscience. All five are essential in some combination or the other to ensure individual success. While the first three would see you through in the immediate level, the latter two are for the longer run. These two establish one as a role model for the men one leads. The beauty is that these two traits are internal; these qualities cannot be cultivated or developed. These two would help one align the individual goals to the organizational goals and at that time, individual success doesn't matter any longer.'

'Sir, the past five years were full of events and milestones. Is there anything you would like to undo?', I asked. 'Nothing,' he said with all his conviction.

He added further, 'In the first year of my tenure as the chief, I made some structural changes in the administrative set-up in DOS. That was essential to bring balance in the system. While I don't regret the decision, I should have executed it better.

'Finally, be aware that you will have to walk, mostly alone, facing the odds that would come all at once. Embrace the opportunity that each of the odds have to offer. This is how you evolve into higher planes of enlightenment and achievement. While you are flying high, many would flock around with their own agendas. Beware of them, at the same time ensure that you identify and invest in your well-wishers.'

I could relate to what he had just said and what I had observed during the last five years of working closely with him. Something flashed in my mind and I asked, 'Have you ever considered writing your biography?'

He smiled and asked, 'Why do you say so?'

'Well, I was just thinking you are superannuating the day after tomorrow. You have a terrific career behind you and an ocean of learning. I feel that it would be a sin if you don't pass on the experiences to us at ISRO and to all who are interested in it. Why don't you give it a serious thought?' I suggested.

'You know, Rajan Sahib also told me long ago, during the GSLV days, that I should write down my experiences,' he recollected.

Our conversation was interrupted as the captain of the aircraft announced that the mist had cleared enough for the aircraft to take-off. He asked the passengers to take their seats as the aircraft got ready to reach the active runway to take-off to Bengaluru.

Dr Radhakrishnan stood up and started to move towards his own seat, and I tried again, 'Sir, if you ever think of writing your biography, I would love to help you on the project in any way I can . . .'

INDEX